SpringerBriefs in Electrical and Computer Engineering

Control, Automation and Robotics

T0183198

Series editors

Tamer Başar, Urbana, USA
Antonio Bicchi, Pisa, Italy
Miroslav Krstic, La Jolla, USA

For further volumes:
http://www.springer.com/series/10198

Agnieszka B. Malinowska
Delfim F. M. Torres

Quantum Variational Calculus

 Springer

Agnieszka B. Malinowska
Department of Mathematics
Bialystok University of Technology,
  Faculty of Computer Science
Bialystok
Poland

Delfim F. M. Torres
Department of Mathematics
University of Aveiro
Aveiro
Portugal

ISSN 2191-8112   ISSN 2191-8120   (electronic)
ISBN 978-3-319-02746-3   ISBN 978-3-319-02747-0   (eBook)
DOI 10.1007/978-3-319-02747-0
Springer Cham Heidelberg New York Dordrecht London

Library of Congress Control Number: 2013951503

Printed on acid-free paper

Springer is part of Springer Science+Business Media (www.springer.com)

To our collaborators and friends,
and to all passionately curious

Preface

Many physical phenomena are described by equations involving nondifferentiable functions, e.g., generic trajectories of quantum mechanics (Feynman and Hibbs 1965). Several different approaches to deal with nondifferentiable functions are proposed in the literature of variational calculus. We can mention the time scale approach, which typically deal with delta or nabla differentiable functions (Ferreira and Torres 2008; Malinowska and Torres 2009; Martins and Torres 2009); the fractional approach, allowing to consider functions that have no first order derivative but have fractional derivatives of all orders less than one (Almeida et al. 2010; Frederico and Torres 2008; Malinowska and Torres 2012); and the quantum approach, which is the subject of this book and is particularly useful to model physical and economical systems (Bangerezako 2004; Cresson et al. 2009; Malinowska and Torres 2010).

Quantum difference operators are receiving an increase of interest, mainly due to their applications—see, e.g., (Almeida and Torres 2009a; Annaby et al. 2012; Bangerezako 2004; Bangerezako 2005; Cresson et al. 2009; Ernst 2008; Kac and Cheung 2002). In 1992, Nottale introduced the theory of scale-relativity without the hypothesis of space–time differentiability (Nottale 1992; Nottale 1999). A rigorous mathematical foundation to Nottale's scale-relativity theory is nowadays given by means of a quantum calculus (Almeida and Torres 2009a; Almeida and Torres 2010; Cresson et al. 2009; Kac and Cheung 2002). Roughly speaking, we substitute the classical derivative by a difference operator, which allows us to deal with sets of nondifferentiable curves. For a deeper discussion of the motivation to study a nondifferentiable quantum calculus and its leading role in the understanding of complex physical systems, we refer the reader to (Almeida and Torres 2009a; Cresson et al. 2009; Kac and Cheung 2002; Nottale 1992).

Quantum calculus has several different dialects (Brito da Cruz et al. 2012; Brito da Cruz et al. 2013b, c; Ernst 2008; Kac and Cheung 2002). The most common one is based on Jackson's q-operators, where q stands for quantum (Annaby and Mansour 2012; Jackson 1908; Jackson 1910; Kac and Cheung 2002). The Jackson q-difference operator is defined by

$$D_q f(t) = \frac{f(qt) - f(t)}{t(q-1)}, \quad t \neq 0,$$

where q is a fixed number, normally taken from $(0, 1)$. Here f is supposed to be defined on a q-geometric set A, i.e., A is a subset of $\mathbb{R}$ (or $\mathbb{C}$) for which $qt \in A$ whenever $t \in A$. The derivative at zero is defined to be $f'(0)$, provided that $f'(0)$ exists (Abu Risha et al. 2007; Andrews et al. 1999; Carmichael 1911; Carmichael 1913; Ismail 2005; Jackson 1908). Jackson also introduced the q-integral

$$\int_0^a f(t) d_q t = a(1 - q) \sum_{k=0}^{\infty} q^k f(aq^k),$$

provided that the series converges, and in this case he defined

$$\int_a^b f(t) d_q t = \int_0^b f(t) d_q t - \int_0^a f(t) d_q t$$

(Al-Salam 1966; Jackson 1908; Jackson 1910; Kac and Cheung 2002). In 1949, Hahn introduced the quantum difference operator

$$D_{q,\omega}[f](t) = \frac{f(qt + \omega) - f(t)}{(q - 1)t + \omega}, \quad t \neq \omega_0,$$

where $\omega_0 := \frac{\omega}{1-q}$, f is a real function defined on an interval I containing ω_0, and $q \in (0, 1)$ and $\omega \geq 0$ are real fixed numbers (Hahn 1949). The Hahn operator unifies (in the limit) the two most well known and used quantum difference operators: the Jackson q-difference derivative D_q, where $q \in (0, 1)$ (Gasper and Rahman 2004; Jackson 1951; Kac and Cheung 2002); and the forward difference Δ_ω, where $\omega > 0$ (Bird 1936; Jagerman 2000; Jordan 1965). The Hahn difference operator is a successful tool for constructing families of orthogonal polynomials and investigating some approximation problems—see, e.g., (Alvarez-Nodarse 2006; Costas-Santos and Marcellán 2007; Dobrogowska and Odzijewicz 2006; Kwon et al. 1998; Petronilho 2007). However, during 60 years, the construction of the proper inverse of Hahn's difference operator $D_{q,\omega}$ remained an open question. Eventually, the problem was solved in 2009 by Aldwoah, who developed the associated integral calculus (Aldwoah 2009)—see also (Aldwoah and Hamza 2011; Annaby et al. 2012). A different approach would be to reduce the Hahn analysis to the Jackson q-analysis (Odzijewicz et al. 2001, Appendix A).

In this book, we develop the variational Hahn calculus. More precisely, we investigate problems of the calculus of variations using Hahn's difference operator and the Jackson–Nörlund integral. The calculus of variations is a classical area of mathematics with many applications in geometry, physics, economics, biology, engineering, dynamical systems, and control theory (Leizarowitz 1985; Leizarowitz 1989; Weinstock 1974). Although being an old theory, it is very much alive and still evolving—see, e.g., (Almeida et al. 2010; Almeida and Torres 2009b; Leizarowitz and Zaslavski 2003; Malinowska and Torres 2012; Martins and Torres 2009). The basic problem of calculus of variations can be formulated as follows: among all differentiable functions $y : [a, b] \to \mathbb{R}$ such that $y(a) = \alpha$ and $y(b) = \beta$, where α, β are fixed real numbers, find the one that minimize (or maximize) the functional

$$\mathcal{L}[y] = \int_a^b L(t, y(t), y'(t)) dt.$$

It can be proved that the candidates to be minimizers (resp. maximizers) to this problem must satisfy the ordinary differential equation

$$\frac{d}{dt} \partial_3 L(t, y(t), y'(t)) = \partial_2 L(t, y(t), y'(t)),$$

called the Euler–Lagrange equation (by $\partial_i L$ we denote the partial derivative of L with respect to its ith argument). If the boundary condition $y(a) = \alpha$ is not present in the problem, then to find the candidates for extremizers one has to add another necessary condition: $\partial_3 L(a, y(a), y'(a)) = 0$; if $y(b) = \beta$ is not present, then $\partial_3 L(b, y(b), y'(b)) = 0$. These two conditions are known as natural boundary conditions or transversality conditions. Since many important physical phenomena are described by nondifferentiable functions, to develop a calculus of variations based on the Hahn quantum operator is an important issue. This is precisely what we do in this book. We discuss the fundamental concepts of a variational calculus, such as the Euler–Lagrange equations for the basic and isoperimetric problems, as well as Lagrange and optimal control problems. As particular cases, we obtain the classical discrete-time calculus of variations (Kelley and Peterson 2001, Chap. 8), the variational q-calculus (Bangerezako 2004; Bangerezako 2005), and the calculus of variations applied to Nörlund's sum (Fort 1937; Fort 1948). Variational functionals that depend on higher-order quantum derivatives are considered as well. Such problems arise in a natural way in applications of engineering, physics, and economics. As an example, we can consider the equilibrium of an elastic bending beam. Let us denote by $y(x)$ the deflection of the point x of the beam, $E(x)$ the elastic stiffness of the material, that can vary with x, and $\xi(x)$ the load that bends the beam. One may assume that, due to some constraints of physical nature, the dynamics does not depend on the usual derivative $y'(x)$ but on some quantum derivative $D_{q,\omega}[y](x)$. In this condition, the equilibrium of the beam correspond to the solution of the following higher-order Hahn's quantum variational problem:

$$\int_0^L \left[\frac{1}{2} \left(E(x) D_{q,\omega}^2 [y](x) \right)^2 - \xi(x) y \left(q^2 x + q\omega + \omega \right) \right] dx \to \min.$$

Note that we recover the classical problem of the equilibrium of the elastic bending beam when $(\omega, q) \to (0, 1)$. This problem is a particular case of problem (P) investigated in Sect. 2.7. Our higher-order Hahn's quantum Euler–Lagrange equation (Theorem 2.45) gives the main tool to solve such problems. As particular cases, we obtain the q-calculus Euler–Lagrange equation (Bangerezako 2004) and the h-calculus Euler–Lagrange equation (Bastos et al. 2011; Kelley and Peterson 2001).

Another generalization of the q-calculus considered in this book includes the quantum calculus that results from the n-power difference operator

$$D_n f(t) = \begin{cases} \frac{f(t^n)-f(t)}{t^n-t} & \text{if } t \in \mathbb{R}\backslash\{-1,0,1\}, \\ f'(t) & \text{if } t \in \{-1,0,1\}, \end{cases}$$

where n is a fixed odd positive integer (Aldwoah 2009). For that we develop a calculus based on the new and more general proposed operator $D_{n,q}$ (see Definition 3.2). The class of quantum systems thus obtained has two parameters and is wider than the standard class of quantum dynamical systems studied in the literature. We claim that the n,q-calculus offers a better mathematical modeling technique to deal with quantum physical systems of time-varying graininess. We trust that our n,q-quantum calculus will become a useful tool to investigate nonconservative dynamical systems in physics (Bartosiewicz and Torres 2008; El-Nabulsi and Torres 2007; El-Nabulsi and Torres 2008; Frederico and Torres 2007).

The subject of this short book is recent and is still evolving. The Hahn quantum variational calculus was started only in 2010 with the work (Malinowska and Torres 2010). Quantum variational problems involving Hahn's derivatives of higher-order were first investigated in (Brito da Cruz et al. 2012). Several quantum variational problems have been recently posed and studied (Aldwoah et al. 2012; Almeida and Torres 2009; Almeida and Torres 2011; Bangerezako 2004; Bangerezako 2005; Brito da Cruz et al. 2013a; Cresson 2005; Cresson et al. 2009; Frederico and Torres 2013; Martins and Torres 2012). The main purpose of this book is to present optimality conditions for generalized quantum variational problems in an unified and a coherent way, and call attention to a promising research area with possible applications in optimal control, physics, and economics (Cruz et al. 2010; Malinowska and Martins 2013; Sengupta 1997). The results presented in the book allow to deal with economical problems with a dynamic nature that does not depend on the usual derivative or the forward difference operator, but on the Hahn quantum difference operator $D_{q,\omega}$. This is connected with a moot question: what kind of "time" (continuous or discrete) should be used in the construction of dynamic models in economics? Although individual economic decisions are generally made at discrete time intervals, it is difficult to believe that they are perfectly synchronized as postulated by discrete models. The usual assumption that the economic activity takes place continuously is a convenient abstraction in many applications. In others, such as the ones studied in financial market equilibrium, the assumption of continuous trading corresponds closely to reality. We believe that our Hahn's approach helps to bridge the gap between two families of models: continuous and discrete.

This short book gives a gentle but solid introduction to the *Quantum Variational Calculus*. The audience is primarily advanced undergraduate and graduate students of mathematics, physics, engineering, and economics. However, the book provides also an opportunity for an introduction to the quantum variational calculus even for experienced researchers. Our aim is to introduce the theory of the quantum calculus of variations in a way suitable for self-study, and at the same time to give the reader the state of the art of a very active and promising research area. We will be extremely happy if the present book will motivate and encourage some readers to follow a research activity in the area, and to take part in the exploration of this exciting subject.

Keywords: Hahn's difference operator; Jackson–Norlünd's integral; Quantum calculus; q-differences; Calculus of variations and optimal control; Quantum variational problems; Necessary optimality conditions; Euler–Lagrange equations; Generalized natural boundary conditions; Isoperimetric problems; Leitmann's principle; Ramsey model; n,q-power difference operator; Generalized Nörlund sum; Generalized Jackson integral; n,q-difference equations.

Bialystok and Aveiro, July 2013   Agnieszka B. Malinowska
                                                                                                  Delfim F. M. Torres

References

Abu Risha MH, Annaby MH, Ismail MEH, Mansour ZS (2007) Linear q-difference equations. Z Anal Anwend 26(4):481–494

Aldwoah KA (2009) Generalized time scales and associated difference equations. Ph.D. Thesis, Cairo University, Cairo

Aldwoah KA, Hamza AE (2011) Difference time scales. Int J Math Stat 9(A11):106–125

Aldwoah KA, Malinowska AB, Torres DFM (2012) The power quantum calculus and variational problems. Dyn Contin Discrete Impuls Syst Ser B Appl Algorithms 19(1–2):93–116

Almeida R, Malinowska AB, Torres DFM (2010) A fractional calculus of variations for multiple integrals with application to vibrating string. J Math Phys 51(3):033503

Almeida R, Torres DFM (2009a) Hölderian variational problems subject to integral constraints. J Math Anal Appl 359(2):674–681

Almeida R, Torres DFM (2009b) Calculus of variations with fractional derivatives and fractional integrals. Appl Math Lett 22(12):1816–1820

Almeida R, Torres DFM (2010) Generalized Euler–Lagrange equations for variational problems with scale derivatives. Lett Math Phys 92(3):221–229

Almeida R, Torres DFM (2011) Nondifferentiable variational principles in terms of a quantum operator. Math Methods Appl Sci 34(18):2231–2241

Al-Salam WA (1966) q-analogues of Cauchy's formulas. Proc Amer Math Soc 17:616–621

Álvarez-Nodarse R (2006) On characterizations of classical polynomials. J Comput Appl Math 196(1):320–337

Andrews GE, Askey R, Roy R (1999) Special functions. Cambridge University Press, Cambridge

Annaby MH, Hamza AE, Aldwoah KA (2012) Hahn difference operator and associated Jackson-Nörlund integrals. J Optim Theory Appl 154(1):133–153

Annaby MH, Mansour ZS (2012) q-fractional calculus and equations. Lecture Notes in Mathematics, 2056, Springer, Heidelberg

Bangerezako G (2004) Variational q-calculus. J Math Anal Appl 289(2):650–665

Bangerezako G (2005) Variational calculus on q-nonuniform lattices. J Math Anal Appl 306(1):161–179

Bartosiewicz Z, Torres DFM (2008) Noether's theorem on time scales. J Math Anal Appl 342(2):1220–1226

Bastos NRO, Ferreira RAC, Torres DFM (2011) Discrete-time fractional variational problems. Sig Process 91(3):513–524

Bird MT (1936) On generalizations of sum formulas of the Euler–Maclaurin type. Amer J Math 58(3):487–503

Brito da Cruz AMC, Martins N, Torres DFM (2012) Higher-order Hahn's quantum variational calculus. Nonlinear Anal 75(3):1147–1157

Brito da Cruz AMC, Martins N, Torres DFM (2013a) Hahn's symmetric quantum variational calculus. Numer Algebra Control Optim 3(1):77–94

Brito da Cruz AMC, Martins N, Torres DFM (2013b) A symmetric quantum calculus. In: Differential and Difference Equations with Applications. Springer Proceedings in Mathematics & Statistics 47. Springer, New York, pp 359–366

Brito da Cruz AMC, Martins N, Torres DFM (2013c) A symmetric Nörlund sum with application to inequalities. In: Differential and Difference Equations with Applications. Springer Proceedings in Mathematics & Statistics 47. Springer, New York, pp 495–503

Carmichael RD (1911) Linear difference equations and their analytic solutions. Trans Amer Math Soc 12(1):99–134

Carmichael RD (1913) On the theory of linear difference equations. Amer J Math 35(2):163–182

Costas-Santos RS, Marcellán F (2007) Second structure relation for q-semiclassical polynomials of the Hahn tableau. J Math Anal Appl 329(1):206–228

Cresson J (2005) Non-differentiable variational principles. J Math Anal Appl 307(1):48–64

Cresson J, Frederico GSF, Torres DFM (2009) Constants of motion for non-differentiable quantum variational problems. Topol Methods Nonlinear Anal 33(2):217–231

Cruz PAF, Torres DFM, Zinober ASI (2010) A non-classical class of variational Problems. Int J Math Model Numer Optim 1(3):227–236

Dobrogowska A, Odzijewicz A (2006) Second order q-difference equations solvable by factorization method. J Comput Appl Math 193(1):319–346

El-Nabulsi RA, Torres DFM (2007) Necessary optimality conditions for fractional action-like integrals of variational calculus with Riemann–Liouville derivatives of order (α, β). Math Methods Appl Sci 30(15):1931–1939

El-Nabulsi RA, Torres DFM (2008) Fractional action like variational problems. J Math Phys 49(5):053521

Ernst T (2008) The different tongues of q-calculus. Proc Est Acad Sci 57(2):81–99

Ernst T (2012) A comprehensive treatment of q-calculus. Birkhäuser/Springer BaselAG, Basel

Ferreira RAC, Torres DFM (2008) Higher-order calculus of variations on time scales. In: Mathematical control theory and finance. Springer, Berlin, pp 149–159

Feynman RP, Hibbs AR (1965) Quantum mechanics and path integrals. McGraw-Hill, New York

Fort T (1937) The calculus of variations applied to Nörlund's sum. Bull Amer Math Soc 43(12):885–887

Fort T (1948) Finite differences and difference equations in the real domain. Clarendon Press, Oxford

Frederico GSF, Torres DFM (2007) A formulation of Noether's theorem for fractional problems of the calculus of variations. J Math Anal Appl 334(2):834–846

Frederico GSF, Torres DFM (2008) Fractional conservation laws in optimal control theory. Nonlinear Dyn 53(3):215–222

Frederico GSF, Torres DFM (2013) A non-differentiable quantum variational embedding in presence of time delays. Int J Differ Equ 8(1):49–62

Gasper G, Rahman M (2004) Basic hypergeometric series, 2nd edn. Cambridge University Press, Cambridge

Hahn W (1949) Über orthogonalpolynome, die q-differenzenlgleichungen genügen. Math Nachr 2:4–34

Ismail MEH (2005) Classical and quantum orthogonal polynomials in one variable. Cambridge University Press, Cambridge

Jackson FH (1908) On q-functions and a certain difference operator. Trans Roy Soc Edinburgh 46:64-72

Jackson FH (1910) On q-definite integrals. Quart J Pure and Appl Math 41:193–203

Jackson FH (1951) Basic integration. Quart J Math, Oxford Ser 2(2):1–16

Jagerman DL (2000) Difference equations with applications to queues. Dekker, NewYork

Jordan C (1965) Calculus of finite differences, 3rd edn. Introduction by Carver, H C Chelsea, New York

Kac V, Cheung P (2002) Quantum calculus. Springer, New York

Kelley WG, Peterson AC (2001) Difference equations, 2nd edn. Harcourt/Academic Press, San Diego, CA

Kwon KH, Lee DW, Park SB, Yoo BH (1998) Hahn class orthogonal polynomials. Kyungpook Math J 38(2):259–281

Leizarowitz A (1985) Infinite horizon autonomous systems with unbounded cost. Appl Math Optim 13(1):19–43

Leizarowitz A (1989) Optimal trajectories of infinite-horizon deterministic control systems. Appl Math Optim 19(1):11–32

Leizarowitz A, Zaslavski AJ (2003) Infinite-horizon discrete-time optimal control problems. J Math Sci (N Y) 116(4):3369–3386

Malinowska AB, Martins N (2013) Generalized transversality conditions for the Hahn quantum variational calculus. Optimization 62(3):323–344

Malinowska AB, Torres DFM (2009) Strong minimizers of the calculus of variations on time scales and the Weierstrass condition. Proc Est Acad Sci 58(4):205–212

Malinowska AB, Torres DFM (2010) The Hahn quantum variational calculus. J Optim Theory Appl 147(3):419–442

Malinowska AB, Torres DFM (2012) Introduction to the fractional calculus of variations. Imperial College Press, London

Martins N, Torres DFM (2009) Calculus of variations on time scales with nabla derivatives. Nonlinear Anal 71(12):e763–e773

Martins N, Torres DFM (2012) Higher-order infinite horizon variational problems in discrete quantum calculus. Comput Math Appl 64(7):2166–2175

Nottale L (1992) The theory of scale relativity. Int J Mod Phys A 7(20):4899–4936

Nottale L (1999) The scale-relativity program. Chaos Solitons Fractals 10(2-3):459–468

Odzijewicz A, Horowski M, Tereszkiewicz A (2001) Integrable multi-boson systems and orthogonal polynomials. J Phys A 34(20):4353–4376

Petronilho J (2007) Generic formulas for the values at the singular points of some special monic classical $H_{q,\omega}$-orthogonal polynomials. J Comput Appl Math 205(1):314–324

Sengupta JK (1997) Recent models in dynamic economics: problems of estimating terminal conditions. Int J Syst Sci 28:857–864

Silva CJ, Torres DFM (2006) Absolute extrema of invariant optimal control problems. Commun Appl Anal 10(4):503–515

Weinstock R (1974) Calculus of variations. With applications to physics and engineering. Reprint of the 1952 edn, Dover, New York

Acknowledgments

This work was supported by *FEDER* funds through *COMPETE*—Operational Programme Factors of Competitiveness ("Programa Operacional Factores de Competitividade") and by Portuguese funds through the *Center for Research and Development in Mathematics and Applications* (University of Aveiro) and the Portuguese Foundation for Science and Technology ("FCT—Fundação para a Ciência e a Tecnologia"), within project PEst-C/MAT/UI4106/2011 with COMPETE number FCOMP-01-0124-FEDER-022690. Malinowska was also supported by Bialystok University of Technology, grant number S/WI/2/11; Torres by EU funding under the 7th Framework Programme FP7-PEOPLE-2010-ITN, grant agreement number 264735-SADCO.

Any comments or suggestions related to the material here contained are more than welcome, and may be submitted by post or by electronic mail to the authors.

Contents

Chapter 1
The Classical Calculus of Variations

For convenience of the reader, we begin with some well known definitions and facts from the classical calculus of variations. Results are given without proofs. For proofs and detailed discussions, we refer the reader to one of the many books on the subject, e.g., van Brunt (2004). For our purposes, the present chapter gives all the necessary background.

The calculus of variations deals with finding extrema and, in this sense, it can be considered a branch of optimization. The problems and techniques in this branch, however, differ markedly from those involving the extrema of functions of several variables owing to the nature of the domain on the quantity to be optimized. The calculus of variations is concerned with finding extrema for functionals, i.e., for mappings from a set of functions to the real numbers. The candidates in the competition for an extremum are thus functions as opposed to vectors in $\mathbb{R}^n$, and this furnishes the subject a distinct character. The functionals are generally defined by definite integrals; the set of functions are often defined by boundary conditions and smoothness requirements, which arise in the formulation of the problem/model. Let us take a look at the classical (basic) problem of the calculus of variations: find a function $y \in C^1([a, b])$ such that

$$\mathcal{L}[y(\cdot)] = \int_a^b L(t, y(t), y'(t))dt \longrightarrow \min, \quad y(a) = y_a, \quad y(b) = y_b, \quad (1.1)$$

with $a, b, y_a, y_b \in \mathbb{R}$ and $L(t, u, v)$ satisfying some smoothness properties.

The enduring interest in the calculus of variations is in part due to its applications. We now present an historical example of this.

Example 1.1 (Brachystochrones). The history of the calculus of variations essentially begins with a problem posed by Johann Bernoulli (1696) as a challenge to the mathematical comunity and in particular to his brother Jacob. The problem is important in the history of the calculus of variations because the method developed by Johann's pupil, Euler, to solve this problem provided a sufficiently general framework to solve other variational problems (van Brunt 2004).

A. B. Malinowska and D. F. M. Torres, *Quantum Variational Calculus*, SpringerBriefs in Control, Automation and Robotics, DOI: 10.1007/978-3-319-02747-0_1, © The Author(s) 2014

The problem that Johann posed was to find the shape of a wire along which a bead initially at rest slides under gravity from one end to the other in minimal time. The endpoints of the wire are specified and the motion of the bead is assumed frictionless. The curve corresponding to the shape of the wire is called a *brachystochrone* or a curve of fastest descent.

The problem attracted the attention of various mathematicians throughout the time including Huygens, L'Hôpital, Leibniz, Newton, Euler and Lagrange (see van Brunt (2004) and references cited therein for more historical details).

To model Bernoulli's problem we use Cartesian coordinates with the positive y-axis oriented in the direction of the gravitational force. Let (a, y_a) and (b, y_b) denote the coordinates of the initial and final positions of the bead, respectively. Here, we require that $a < b$ and $y_a < y_b$. The problem consists of determining, among the curves that have (a, y_a) and (b, y_b) as endpoints, the curve on which the bead slides down from (a, y_a) to (b, y_b) in minimum time. The problem makes sense only for continuous curves. We make the additional simplifying (but reasonable) assumptions that the curve can be represented by a function $y : [a, b] \to \mathbb{R}$ and that y is at least piecewise differentiable in the interval $[a, b]$. The total time it takes the bead to slide down a curve is given by

$$T[y(\cdot)] = \int_0^l \frac{ds}{v(s)}, \tag{1.2}$$

where l denotes the arclength of the curve, s is the arclength parameter, and v is the velocity of the bead s units down the curve from (a, b).

We now derive an expression for the velocity in terms of the function y. We use the law of conservation of energy to achieve this. At any position $(x, y(x))$ on the curve, the sum of the potential and kinetic energies of the bead is a constant. Hence

$$\frac{1}{2}mv^2(x) + mgy(x) = c, \tag{1.3}$$

where m is the mass of the bead, v is the velocity of the bead at $(x, y(x))$, and c is a constant. Solving equation (1.3) for v gives

$$v(x) = \sqrt{\frac{2c}{m} - 2gy(x)}.$$

Equality (1.2) becomes

$$T[y(\cdot)] = \int_a^b \frac{\sqrt{1 + y'^2(x)}}{\sqrt{\frac{2c}{m} - 2gy(x)}} dx.$$

We thus seek a function y such that T is minimum and $y(a) = y_a$, $y(b) = y_b$.

It can be shown that the extrema for T is a portion of the curve called *cycloid* (cf. Example 2.3.4 in van Brunt (2004)).

1.1 Problem Statement

The calculus of variations is concerned with the problem of finding minima (or maxima) of a functional $\mathcal{J} : \mathcal{D} \to \mathbb{R}$, where $\mathcal{D}$ is a subset of a (normed) linear space $\mathbf{D}$ of real-valued (or real-vector-valued) functions. The formulation of a problem requires two steps: the specification of a performance criterion, and the statement of physical constraints that should be satisfied. The performance criterion $\mathcal{J}$, also called cost functional (or objective), must be specified for evaluating quantitatively the performance of the system under study. The typical form of the cost is:

$$\mathcal{J}(y) = \int_a^b L(t, y(t), y'(t)) \, dt,$$

where $t \in [a, b]$ is the independent variable, usually called time; $y(t) \in \mathbb{R}^N, N \geq 1$, is a real vector variable, the functions $y(t), a \leq t \leq b$, are generally called trajectories or curves; $y'(t) \in \mathbb{R}^N$ stands for the derivative of $y(t)$ with respect to time t; and $L : [a, b] \times \mathbb{R}^{2N} \to \mathbb{R}$ is a real-valued function, called the Lagrangian.

Enforcing constraints in the optimization problem reduces the set of candidate functions and leads to the following definition.

Definition 1.2 A trajectory $y \in \mathbf{D}$ is said to be an admissible trajectory (or admissible function), provided it satisfies all the constraints of the problem along the interval $[a, b]$. The set of admissible trajectories is denoted by $\mathcal{D}$.

A great variety of constraints is considered. The simplest one are boundary conditions, e.g., $y(a) = y_a$ and $y(b) = y_b$, $y_a, y_b \in \mathbb{R}^N$ or we may require that the trajectory $y \in \mathbf{D}$ join a fixed point (a, y_a) to a specified curve $f(t), a \leq t \leq T$. Besides boundary constraints, another type of constraints can be considered,

$$\mathcal{G}^j(y) = \int_a^b G^j(t, y(t), y'(t)) dt = l_j, \quad j = 1, \ldots, r, \quad r \geq 1,$$

where $G^j : [a, b] \times \mathbb{R}^{2N} \to \mathbb{R}, j = 1, \ldots, r$. These constraints are often referred to as isoperimetric constraints. More generally, constraints of the form

$$G^j(t, y(t), y'(t)) = 0, \quad j = 1, \ldots, r, \quad r \geq 1,$$

are called constraints of Lagrange form.

Having defined an objective functional $\mathcal{J}$ and constraints, one must then decide about the class of functions with respect to which the optimization shall be performed. The traditional choice in the calculus of variations is to consider the class of continuously differentiable functions, e.g., $C^1([a, b])$. We endow $C^1([a, b])$ with a norm. The most natural choice for a norm on $C^1([a, b])$ is $\|y\|_{1,\infty} := \max_{a \leq t \leq b} \|y(t)\| + \max_{a \leq t \leq b} \|y'(t)\|$, where $\| \cdot \|$ stands for the Euclidean norm in $\mathbb{R}^N$. The class of functions $C^1([a, b])$ endowed with $\| \cdot \|_{1,\infty}$ is a Banach space.

Let us now define what is meant by a minimum of $\mathcal{J}$ on $\mathcal{D}$.

Definition 1.3 A trajectory $\bar{y} \in \mathcal{D}$ is said to be a local minimizer (resp. local maximizer) for $\mathcal{J}$ on $\mathcal{D}$, if there exists $\delta > 0$ such that $\mathcal{J}(\bar{y}) \leq \mathcal{J}(y)$ (resp. $\mathcal{J}(\bar{y}) \geq \mathcal{J}(y)$) for all $y \in \mathcal{D}$ with $\|y - \bar{y}\|_{1,\infty} < \delta$.

The concept of variation of a functional is central to the solution of problems of the calculus of variations.

Definition 1.4 The first variation of $\mathcal{J}$ at $y \in \mathbf{D}$ in the direction $y \in \mathbf{D}$ is defined as

$$\delta \mathcal{J}(y; h) := \lim_{\varepsilon \to 0} \frac{\mathcal{J}(y + \varepsilon h) - \mathcal{J}(y)}{\varepsilon} = \frac{\partial}{\partial \varepsilon} \mathcal{J}(y + \varepsilon h)\Big|_{\varepsilon=0} ,$$

provided the limit exists.

Definition 1.5 A direction $h \in \mathbf{D}$, $h \neq 0$, is said to be an admissible variation for $\mathcal{J}$ at $y \in \mathcal{D}$ if

(i) $\delta \mathcal{J}(y; h)$ exists; and
(ii) $y + \varepsilon h \in \mathcal{D}$ for all sufficiently small ε.

The following well known result provides a necessary optimality condition for the problems of the calculus of variations, based on the concept of variation.

Theorem 1.6 *Let $\mathcal{J}$ be a functional defined on $\mathcal{D}$. Suppose that y is a local minimizer (or local maximizer) for $\mathcal{J}$ on $\mathcal{D}$. Then, $\delta \mathcal{J}(y; h) = 0$ for each admissible variation h at y.*

1.2 The Euler–Lagrange Equations

In this section, we present a first-order necessary optimality condition for a problem which is know as the elementary (or basic or fundamental) problem of the calculus of variations.

The next lemma is an essential result upon which the calculus of variations depends. It is called the fundamental lemma of the calculus of variations, sometimes also called the DuBois–Reymond lemma.

Lemma 1.7 (The fundamental lemma of the calculus of variations). *If $g(t)$ is a continuous function of t for $a \leq t \leq b$, and if*

$$\int_a^b g(t)h(t)\,dt = 0$$

for all functions $h(t)$ that are continuous for $a \leq t \leq b$ and are zero at $t = a$ and $t = b$, then $g(t) = 0$ for all $a \leq t \leq b$.

We denote by $\partial_i K$, $i = 1, \ldots, M$ ($M \in \mathbb{N}$), the partial derivative of a function $K : \mathbb{R}^M \to \mathbb{R}$ with respect to its ith argument. The following theorem gives a necessary optimality condition for the fundamental problem of the calculus of variations.

Theorem 1.8 (The Euler–Lagrange equations). *Consider the problem of minimizing (or maximizing) the functional*

$$\mathcal{J}(y) = \int_a^b L(t, y(t), y'(t)) \, dt$$

on $\mathcal{D} = \{y \in \mathbf{D} : y(a) = y_a, \ y(b) = y_b\}$, where $L : [a, b] \times \mathbb{R}^{2N} \to \mathbb{R}$ is a continuously differentiable function. Suppose that y gives a (local) minimum (or maximum) to $\mathcal{J}$ on $\mathcal{D}$. Then,

$$\partial_i L(t, y(t), y'(t)) = \frac{d}{dt} \partial_{N+i} L(t, y(t), y'(t)), \quad i = 2, \ldots N + 1, \tag{1.4}$$

for all $t \in [a, b]$.

Definition 1.9 A function y that satisfies the system of Euler–Lagrange equations (1.4) on $[a, b]$ is called an *extremal* for the functional $\mathcal{J}$.

If one of the boundary conditions $y(a) = y_a$ or $y(b) = y_b$ is not present in the problem (it is possible that all are not present), then in order to find extremizers we must add other necessary conditions, usually called the natural boundary conditions (or transversality conditions).

Theorem 1.10 (Natural boundary conditions). *If y is a local minimizer (or maximizer) to the functional*

$$\mathcal{J}(y) = \int_a^b L(t, y(t), y'(t)) \, dt,$$

then y satisfies the Euler–Lagrange equations (1.4). Moreover,

- *if $y(a) = y_a$ is free, then the natural boundary conditions*

$$\partial_{N+i} L(a, y(a), y'(a)) = 0, \quad i = 2, \ldots N + 1, \tag{1.5}$$

  hold;
- *if $y(b)$ is free, then the natural boundary conditions*

$$\partial_{N+i} L(b, y(b), y'(b)) = 0, \quad i = 2, \ldots N + 1, \tag{1.6}$$

  hold.

1.3 Problems with Isoperimetric Constraints

An isoperimetric problem of the calculus of variations is a problem wherein one or more constraints involve the integral of a given function over part or all of the integration horizon $[a, b]$. One of the earliest problem involving such a constraint is that of finding the geometric figure with the largest area that can be enclosed by a curve of some specified length—the famous Queen Dido isoperimetric problem. The following theorems provide a characterization of the extremals for isoperimetric problems, based on the method of Lagrange multipliers.

Theorem 1.11 *Consider the problem of minimizing (or maximizing) the functional*

$$\mathcal{J}(y) = \int_a^b L(t, y(t), y'(t)) \, dt$$

on $\mathcal{D}$ given by those $y \in \mathbf{D}$ such that $y(a) = y_a$, $y(b) = y_b$, and

$$\mathcal{G}(y) = \int_a^b G(t, y(t), y'(t)) dt = l, \tag{1.7}$$

where $L, G : [a, b] \times \mathbb{R}^{2N} \to \mathbb{R}$ are continuously differentiable functions. Suppose that y gives a (local) minimum (or maximum) to this problem. Assume that $\delta \mathcal{G}(y; h)$ does not vanish for all $h \in \mathbf{D}$. Then there exists a constant $\lambda \in \mathbb{R}$ such that y is a solution of the Euler–Lagrange equations

$$\partial_i F(t, y(t), y'(t)) = \frac{d}{dt} \partial_{N+i} F(t, y(t), y'(t)), \quad i = 2, \ldots N+1,$$

where $F(t, y, y', \lambda) = L(t, y, y') - \lambda G(t, y, y')$.

Remark 1.12 The equality (1.7) is called an isoperimetric constraint. Observe that $\delta \mathcal{G}(y; h)$ does not vanish for all $h \in \mathbf{D}$ if and only if y is not an extremal for $\mathcal{G}$.

1.4 Sufficient Optimality Conditions via Joint Convexity

In this section we present a sufficient condition for an extremal to be a global extremizer (minimizer or maximizer).

Definition 1.13 Given a function $f \in C^1([a, b] \times \mathbb{R}^{2N}; \mathbb{R})$, we say that $f(\underline{x}, y, v)$ is jointly convex (resp. jointly concave) in (y, v), if

$$f(x, y + y^0, v + v^0) - f(x, y, v)$$

$$\geq (\leq) \sum_{i=2}^{N+1} \partial_i f(x, y, v) y_{i-1}^0 + \sum_{i=2}^{N+1} \partial_{N+i} f(x, y, v) v_{i-1}^0$$

for all $(x, y, v), (x, y + y^0, v + v^0) \in [a, b] \times \mathbb{R}^{2N}$.

Theorem 1.14 *Let $L(\underline{x}, y, v)$ be jointly convex (resp. jointly concave) in (y, v). If y satisfies the system of N Euler–Lagrange equations (1.4), then y is a global minimizer (resp. global maximizer) to*

$$\mathcal{J}(y) = \int_a^b L(t, y(t), y'(t)) \, dt$$

on $\mathcal{D} = \{y \in \mathbf{D} : y(a) = y_a, y(b) = y_b\}$.

Reference

van Brunt B (2004) The calculus of variations. Springer, New York

Chapter 2
The Hahn Quantum Variational Calculus

We introduce the Hahn quantum variational calculus. Necessary and sufficient optimality conditions for the basic, isoperimetric, and Hahn quantum Lagrange problems, are studied. We also show the validity of Leitmann's direct method (Almeida and Torres 2010b; Carlson and Leitmann 2005a,b, 2008; Leitmann 2002, 2003) for the Hahn quantum variational calculus, and give explicit solutions to some concrete problems. Next, we prove a necessary optimality condition of Euler–Lagrange type for quantum variational problems involving Hahn's derivatives of higher-order. Finally, we extend the previous results and obtain optimality conditions for generalized quantum variational problems with a Lagrangian depending on the free endpoints. To illustrate the results, we provide several examples and discuss quantum versions of the Ramsey model and an adjustment model in economics.

2.1 Preliminaries

Let $q \in]0, 1[$ and $\omega \geq 0$. Define $\omega_0 := \dfrac{\omega}{1 - q}$ and let I be a real interval containing ω_0. For a function f defined on I, the *Hahn difference operator* of f is given by

$$D_{q,\omega}[f](t) := \begin{cases} \dfrac{f(qt + \omega) - f(t)}{(q - 1)t + \omega} & \text{if } t \neq \omega_0 \\ \\ f'(\omega_0) & \text{if } t = \omega_0 \end{cases}$$

provided that f is differentiable at ω_0 (where f' denotes the Fréchet derivative of f). $D_{q,\omega}[f]$ is called the q, ω-*derivative* of f, and f is said to be q, ω-*differentiable on* I if $D_{q,\omega}[f](\omega_0)$ exists.

A. B. Malinowska and D. F. M. Torres, *Quantum Variational Calculus*,
SpringerBriefs in Control, Automation and Robotics,
DOI: 10.1007/978-3-319-02747-0_2, © The Author(s) 2014

Remark 2.1 Note that when $q \to 1$ we obtain the forward h-difference operator

$$\Delta_h [f] (t) := \frac{f(t+h) - f(t)}{h},$$

and when $\omega = 0$ we obtain the Jackson q-difference operator

$$D_{q,0}[f](t) := \begin{cases} \dfrac{f(qt) - f(t)}{(q-1)t} & \text{if } t \neq 0 \\[2mm] f'(0) & \text{if } t = 0 \end{cases}$$

provided $f'(0)$ exists. Hence, we can state that the $D_{q,\omega}$ operator generalizes the forward h-difference and the Jackson q-difference operators (Ernst 2008; Kac and Cheung 2002; Koornwinder 1994). Notice also that, under appropriate conditions,

$$\lim_{q \to 1} D_{q,0} [f] (t) = f'(t).$$

Example 2.2 Let $q = \omega = 1/2$. In this case $\omega_0 = 1$. It is easy to see that $f:[-1, 1] \to \mathbb{R}$ given by

$$f(t) = \begin{cases} -t & \text{if } t \in]-1, 0[\cup]0, 1] \\ 0 & \text{if } t = -1 \\ 1 & \text{if } t = 0 \end{cases}$$

is not a continuous function but is q, ω-differentiable in $[-1, 1]$ with

$$D_{q,\omega}[f](t) = \begin{cases} -1 & \text{if } t \in]-1, 0[\cup]0, 1] \\ 1 & \text{if } t = -1 \\ -3 & \text{if } t = 0. \end{cases}$$

Example 2.3 Let $q \in]0, 1[$, $\omega = 0$, and

$$f(t) = \begin{cases} t^2 & \text{if } t \in \mathbb{Q} \\ -t^2 & \text{if } t \in \mathbb{R} \setminus \mathbb{Q}. \end{cases}$$

Note that f is only Fréchet differentiable in zero, but since $\omega_0 = 0$, f is q, ω-differentiable on the entire real line.

The Hahn difference operator has the following properties:

Theorem 2.4 (Aldwoah 2009; Aldwoah and Hamza 2011) *If $f, g : I \to \mathbb{R}$ are q, ω-differentiable and $t \in I$, then:*

1. $D_{q,\omega}[f](t) \equiv 0$ *on I if and only if f is constant;*
2. $D_{q,\omega}[f+g](t) = D_{q,\omega}[f](t) + D_{q,\omega}[g](t)$;
3. $D_{q,\omega}[fg](t) = D_{q,\omega}[f](t)g(t) + f(qt+\omega)D_{q,\omega}[g](t)$;
4. $D_{q,\omega}\left[\dfrac{f}{g}\right](t) = \dfrac{D_{q,\omega}[f](t)g(t) - f(t)D_{q,\omega}[g](t)}{g(t)g(qt+\omega)}$ *if $g(t)g(qt+\omega) \neq 0$;*
5. $f(qt+\omega) = f(t) + (t(q-1)+\omega)D_{q,\omega}[f](t)$.

Proposition 2.5 (Aldwoah 2009) *Let $a, b \in \mathbb{R}$. We have*

$$D_{q,\omega}(at+b)^n = a\sum_{k=0}^{n-1}(a(qt+\omega)+b)^k(at+b)^{n-k-1},$$

for $n \in \mathbb{N}$ and $t \neq \omega_0$.

Let $\sigma(t) = qt + \omega$, for all $t \in I$. Note that σ is a contraction, $\sigma(I) \subseteq I$, $\sigma(t) < t$ for $t > \omega_0$, $\sigma(t) > t$ for $t < \omega_0$, and $\sigma(\omega_0) = \omega_0$.

We use the following standard notation of q-calculus: for $k \in \mathbb{N}_0 := \mathbb{N} \cup \{0\}$, $[k]_q := \dfrac{1-q^k}{1-q}$.

Lemma 2.6 (Aldwoah 2009) *Let $k \in \mathbb{N}$ and $t \in I$. Then,*

1. $\sigma^k(t) = \underbrace{\sigma \circ \sigma \circ \cdots \circ \sigma}_{k\text{-times}}(t) = q^k t + \omega[k]_q$;

2. $\left(\sigma^k(t)\right)^{-1} = \sigma^{-k}(t) = \dfrac{t - \omega[k]_q}{q^k}$.

Following (Aldwoah 2009; Aldwoah and Hamza 2011) we define the notion of q, ω-*integral* (also known as the *Jackson–Nörlund integral*) as follows:

Definition 2.7 Let $a, b \in I$ and $a < b$. For $f : I \to \mathbb{R}$ the q, ω-*integral of f* from a to b is given by

$$\int_a^b f(t)\, d_{q,\omega}t := \int_{\omega_0}^b f(t)\, d_{q,\omega}t - \int_{\omega_0}^a f(t)\, d_{q,\omega}t,$$

where

$$\int_{\omega_0}^x f(t)\, d_{q,\omega}t := (x(1-q)-\omega)\sum_{k=0}^{+\infty}q^k f\left(xq^k + \omega[k]_q\right), \; x \in I,$$

provided that the series converges at $x = a$ and $x = b$. In that case, f is called q, ω-*integrable on $[a, b]$*. We say that f is q, ω-*integrable over I* if it is q, ω-integrable over $[a, b]$ for all $a, b \in I$.

Remark 2.8 The q, ω-*integral* generalizes the Jackson q-integral and the Nörlund sum (Kac and Cheung 2002). When $\omega = 0$, we obtain the Jackson q-integral

$$\int_a^b f(t)\, d_q t := \int_0^b f(t)\, d_q t - \int_0^a f(t)\, d_q t,$$

where

$$\int_0^x f(t)\, d_q t := x(1-q) \sum_{k=0}^{+\infty} q^k f\left(xq^k\right).$$

When $q \to 1$, we obtain the Nörlund sum

$$\int_a^b f(t)\, \Delta_\omega t := \int_{+\infty}^b f(t)\, \Delta_\omega t - \int_{+\infty}^a f(t)\, \Delta_\omega t,$$

where

$$\int_{+\infty}^x f(t)\, \Delta_\omega t := -\omega \sum_{k=0}^{+\infty} f(x + k\omega).$$

It can be shown that if $f : I \to \mathbb{R}$ is continuous at ω_0, then f is q, ω-integrable over I (see Aldwoah (2009); Aldwoah and Hamza (2011) for the proof).

Theorem 2.9 (Aldwoah 2009) (***Fundamental Theorem of Hahn's Calculus***) *Assume that $f : I \to \mathbb{R}$ is continuous at ω_0 and, for each $x \in I$, define*

$$F(x) := \int_{\omega_0}^x f(t)\, d_{q,\omega} t.$$

Then F is continuous at ω_0. Furthermore, $D_{q,\omega}[F](x)$ exists for every $x \in I$ and $D_{q,\omega}[F](x) = f(x)$. Conversely, $\int_a^b D_{q,\omega}[f](t)\, d_{q,\omega} t = f(b) - f(a)$ for all $a, b \in I$.

Aldwoah proved that the q, ω-*integral* has the following properties:

Theorem 2.10 (Aldwoah 2009; Aldwoah and Hamza 2011) *Let $f, g : I \to \mathbb{R}$ be q, ω-integrable on I, $a, b, c \in I$ and $k \in \mathbb{R}$. Then,*

1. $\displaystyle \int_a^a f(t)\, d_{q,\omega} t = 0;$
2. $\displaystyle \int_a^b kf(t)\, d_{q,\omega} t = k \int_a^b f(t)\, d_{q,\omega} t;$
3. $\displaystyle \int_a^b f(t)\, d_{q,\omega} t = -\int_b^a f(t)\, d_{q,\omega} t;$
4. $\displaystyle \int_a^b f(t)\, d_{q,\omega} t = \int_a^c f(t)\, d_{q,\omega} t + \int_c^b f(t)\, d_{q,\omega} t;$
5. $\displaystyle \int_a^b (f(t) + g(t))\, d_{q,\omega} t = \int_a^b f(t)\, d_{q,\omega} t + \int_a^b g(t)\, d_{q,\omega} t;$
6. *Every Riemann integrable function f on I is q, ω-integrable on I;*

7. *If $f, g : I \to \mathbb{R}$ are q, ω-differentiable and $a, b \in I$, then*

$$\int_a^b f(t) D_{q,\omega}[g](t) d_{q,\omega}t = \left[f(t) g(t) \right]_a^b - \int_a^b D_{q,\omega}[f](t) g(qt + \omega) d_{q,\omega}t.$$

Property 7 of Theorem 2.10 is known as *q, ω-integration by parts formula*.

Lemma 2.11 (Annaby et al. 2012) *Let $s \in I$ and f and g be q, ω-integrable over I. Suppose that*

$$|f(t)| \le g(t), \quad \forall t \in \{q^n s + \omega [n]_q : n \in \mathbb{N}_0\}.$$

If $\omega_0 \le s$, then for $b \in \{q^n s + \omega [n]_q : n \in \mathbb{N}_0\}$

$$\left| \int_{\omega_0}^b f(t) d_{q,\omega}t \right| \le \int_{\omega_0}^b g(t) d_{q,\omega}t.$$

Remark 2.12 Note that there is an inconsistency in Aldwoah (2009). Indeed, Lemma 6.2.7 of Aldwoah (2009) is only valid if $b \ge \omega_0$ and $a \le b$.

Remark 2.13 In general, the Jackson–Nörlund integral does not satisfies the following inequality (for a counterexample see Aldwoah (2009)):

$$\left| \int_a^b f(t) d_{q,\omega}t \right| \le \int_a^b |f(t)| d_{q,\omega}t, \quad a, b \in I.$$

For $s \in I$ we define

$$[s]_{q,\omega} := \{q^n s + \omega [n]_q : n \in \mathbb{N}_0\} \cup \{\omega_0\}.$$

The following definition and lemma are important for our purposes.

Definition 2.14 Let $s \in I$ and $g : I \times] - \bar{\theta}, \bar{\theta}[\to \mathbb{R}$. We say that $g(t, \cdot)$ is differentiable at θ_0 uniformly in $[s]_{q,\omega}$ if for every $\varepsilon > 0$ there exists $\delta > 0$ such that

$$0 < |\theta - \theta_0| < \delta \Rightarrow \left| \frac{g(t, \theta) - g(t, \theta_0)}{\theta - \theta_0} - \partial_2 g(t, \theta_0) \right| < \varepsilon$$

for all $t \in [s]_{q,\omega}$, where $\partial_2 g = \frac{\partial g}{\partial \theta}$.

Lemma 2.15 *Let $s \in I$ and assume that $g : I \times] - \bar{\theta}, \bar{\theta}[\to \mathbb{R}$ is differentiable at θ_0 uniformly in $[s]_{q,\omega}$, $G(\theta) := \int_{\omega_0}^s g(t, \theta) d_{q,\omega}t$ for θ near θ_0, and*

$$\int_{\omega_0}^{s} \partial_2 g\, (t, \theta_0)\, d_{q,\omega} t \quad exist. \ Then, \ G\,(\theta) \ is \ differentiable \ at \ \theta_0 \ with \ G'\,(\theta_0)$$

$$= \int_{\omega_0}^{s} \partial_2 g\, (t, \theta_0)\, d_{q,\omega} t.$$

Proof For $s = \omega_0$ the result is clear. Let $s \neq \omega_0$ and $\varepsilon > 0$ be arbitrary. Since $g(t, \cdot)$ is differentiable at θ_0, uniformly in t, there exists $\delta > 0$, such that, for all $t \in [s]_{q,\omega}$, and for $0 < |\theta - \theta_0| < \delta$, the following inequality holds:

$$\left| \frac{g(t, \theta) - g(t, \theta_0)}{\theta - \theta_0} - \partial_2 g(t, \theta_0) \right| < \frac{\varepsilon}{s - \omega_0}.$$

Applying Theorem 2.10 and Lemma 2.11, for $0 < |\theta - \theta_0| < \delta$, we have

$$\left| \frac{G(\theta) - G(\theta_0)}{\theta - \theta_0} - G'(\theta_0) \right|$$

$$= \left| \frac{\int_{\omega_0}^{s} g(t, \theta) d_{q,\omega} t - \int_{\omega_0}^{s} g(t, \theta_0) d_{q,\omega} t}{\theta - \theta_0} - \int_{\omega_0}^{s} \partial_2 g(t, \theta_0) d_{q,\omega} t \right|$$

$$= \left| \int_{\omega_0}^{s} \left[\frac{g(t, \theta) - g(t, \theta_0)}{\theta - \theta_0} - \partial_2 g(t, \theta_0) \right] d_{q,\omega} t \right|$$

$$< \int_{\omega_0}^{s} \frac{\varepsilon}{s - \omega_0} d_{q,\omega} t = \frac{\varepsilon}{s - \omega_0} \int_{\omega_0}^{s} 1 d_{q,\omega} t = \varepsilon.$$

Hence, $G(\cdot)$ is differentiable at θ_0 and $G'(\theta_0) = \int_{\omega_0}^{s} \partial_2 g(t, \theta_0) d_{q,\omega} t$.

Let $a, b \in I$ with $a < b$. Recall that I is an interval containing ω_0. We define the q, ω-interval by

$$[a, b]_{q,\omega} := \{q^n a + \omega[n]_q : n \in \mathbb{N}_0\} \cup \{q^n b + \omega[n]_q : n \in \mathbb{N}_0\} \cup \{\omega_0\},$$

that is, $[a, b]_{q,\omega} = [a]_{q,\omega} \cup [b]_{q,\omega}$. For $r \in \mathbb{N}$ we introduce the linear space $\mathcal{Y}^r = \mathcal{Y}^r\,([a, b], \mathbb{R})$ by

$$\mathcal{Y}^r := \left\{ y : [a, b] \rightarrow \mathbb{R} \mid D_{q,\omega}^i [y], i = 0, \ldots, r, \right.$$

$$\left. \text{are bounded on } [a, b] \text{ and continuous at } \omega_0 \right\}$$

endowed with the norm

$$\|y\|_{r,\infty} := \sum_{i=0}^{r} \left\| D_{q,\omega}^i [y] \right\|_{\infty},$$

where $\|y\|_{\infty} := \sup_{t \in [a,b]} |y(t)|$.

2.2 The Hahn Quantum Euler–Lagrange Equation

In this section we obtain the Euler–Lagrange equation for the basic problem of the Hahn quantum variational calculus. As in the classical case, we need the following lemma.

Lemma 2.16 (Fundamental Lemma of the Hahn quantum variational calculus) *Let $f \in \mathcal{Y}^0$. One has $\int_a^b f(t)h(qt + \omega)d_{q,\omega}t = 0$ for all functions $h \in \mathcal{Y}^0$ with $h(a) = h(b) = 0$ if and only if $f(t) = 0$ for all $t \in [a,b]_{q,\omega}$.*

Proof The implication "$\Leftarrow$" is obvious. Let us prove the implication "$\Rightarrow$". Suppose, by contradiction, that $f(p) \neq 0$ for some $p \in [a,b]_{q,\omega}$.
Case I If $p \neq \omega_0$, then $p = q^k a + \omega[k]_q$ or $p = q^k b + \omega[k]_q$ for some $k \in \mathbb{N}_0$. Observe that $a(1-q) - \omega$ and $b(1-q) - \omega$ cannot vanish simultaneously. Therefore, without loss of generality, we can assume $a(1-q) - \omega \neq 0$ and $p = q^k a + \omega[k]_q$. Define

$$h(t) = \begin{cases} f(q^k a + \omega[k]_q), & \text{if } t = q^{k+1} a + \omega[k+1]_q \\ 0, & \text{otherwise.} \end{cases}$$

Then,

$$\int_a^b f(t)h(qt + \omega)d_{q,\omega}t$$
$$= -(a(1-q) - \omega)q^k f(q^k a + \omega[k]_q)h(q^{k+1}a + \omega[k+1]_q) \neq 0,$$

which is a contradiction.
Case II If $p = \omega_0$, then without loss of generality we can assume $f(\omega_0) > 0$. We know that (see Aldwoah (2009); Annaby et al. (2012) for more details)

$$\lim_{n\to\infty} q^n a + [n]_{q,\omega} = \lim_{n\to\infty} q^n b + \omega[n]_q = \omega_0.$$

As f is continuous at ω_0, we have

$$\lim_{n\to\infty} f(q^n a + \omega[n]_q) = \lim_{n\to\infty} f(q^n b + \omega[n]_q) = f(\omega_0).$$

Therefore, there exists $N \in \mathbb{N}$, such that for all $n > N$ the inequalities

$$f(q^n a + \omega[n]_q) > 0 \quad \text{and} \quad f(q^n b + \omega[n]_q) > 0$$

hold. If $\omega_0 \neq a, b$, then we define

$$h(t) = \begin{cases} f(q^n b + w[n]_q), & \text{if } t = q^{n+1}a + w[n+1]_q, & \text{for all } n > N \\ f(q^n a + w[n]_q), & \text{if } t = q^{n+1}b + w[n+1]_q, & \text{for all } n > N \\ 0, & \text{otherwise.} \end{cases}$$

Hence,

$$\int_a^b f(t)h(qt+w)d_{q,w}t = (b-a)(1-q)\sum_{n=N}^{\infty} q^n f(q^n a + w[n]_q) f(q^n b + w[n]_q) \neq 0,$$

which is a contradiction. If $w_0 = b$, then we define

$$h(t) = \begin{cases} f(w_0), & \text{if } t = q^{n+1}a + w[n+1]_q, & \text{for all } n > N \\ 0, & \text{otherwise.} \end{cases}$$

Hence,

$$\int_a^b f(t)h(qt + w)d_{q,w}t = -\int_{w_0}^a f(t)h(qt + w)d_{q,w}t$$

$$= -(a(1-q) - w)\sum_{n=N}^{\infty} q^n f(q^n a + w[n]_q) f(w_0) \neq 0,$$

which is a contradiction. Similarly, we show the case when $w_0 = a$.

Consider the following q, w-variational problem

$$\mathcal{L}[y] = \int_a^b L\left(t, y\left(qt + w\right), D_{q,w}\left[y\right](t)\right) d_{q,w}t \longrightarrow \text{extr} \tag{2.1}$$

where "extr" denotes "extremize", in the class of functions $y \in \mathcal{Y}^1$ satisfying the boundary conditions

$$y(a) = \alpha \quad \text{and} \quad y(b) = \beta \tag{2.2}$$

for some fixed $\alpha, \beta \in \mathbb{R}$.

Definition 2.17 A function $y \in \mathcal{Y}^1$ is said to be admissible for (2.1)–(2.2) if it satisfies the endpoint conditions (2.2). We say that $h \in \mathcal{Y}^1$ is an admissible variation for (2.1)–(2.2) if $h(a) = h(b) = 0$.

In the sequel we assume that the Lagrangian L satisfies the following hypotheses:

(H1) $(u_0, u_1) \rightarrow L(t, u_0, u_1)$ is a $C^1(\mathbb{R}^2, \mathbb{R})$ function for any $t \in [a, b]$;
(H2) $t \rightarrow L(t, y(qt + w), D_{q,w}[y](t))$ is continuous at w_0 for any $y \in \mathcal{Y}^1$;
(H3) functions $t \rightarrow \partial_i L(t, y(qt + w), D_{q,w}[y](t)), i = 2, 3$, belong to $\mathcal{Y}^1$ for all $y \in \mathcal{Y}^1$.

Definition 2.18 We say that y_* is a local minimizer (resp. local maximizer) for problem (2.1)–(2.2) if y_* is an admissible function and there exists $\delta > 0$ such that

$$\mathcal{L}[y_*] \le \mathcal{L}[y] \quad (\text{resp. } \mathcal{L}[y_*] \ge \mathcal{L}[y])$$

for all admissible y with $\|y_* - y\|_{1,\infty} < \delta$.

For fixed $y, h \in \mathcal{Y}^1$, we define the real function ϕ by

$$\phi(\varepsilon) := \mathcal{L}[y + \varepsilon h].$$

The first variation for problem (2.1) is defined by

$$\delta\mathcal{L}[y, h] := \phi'(0).$$

Observe that,

$$\mathcal{L}[y + \varepsilon h] = \int_a^b L(t, y(qt + \omega) + \varepsilon h(qt + \omega), D_{q,\omega}[y](t) + \varepsilon D_{q,\omega}[h](t))d_{q,\omega}t$$

$$= \int_{\omega_0}^b L(t, y(qt + \omega) + \varepsilon h(qt + \omega), D_{q,\omega}[y](t) + \varepsilon D_{q,\omega}[h](t))d_{q,\omega}t$$

$$- \int_{\omega_0}^a L(t, y(qt + \omega) + \varepsilon h(qt + \omega), D_{q,\omega}[y](t) + \varepsilon D_{q,\omega}[h](t))d_{q,\omega}t.$$

Writing

$$\mathcal{L}_b[y + \varepsilon h] = \int_{\omega_0}^b L(t, y(qt + \omega) + \varepsilon h(qt + \omega), D_{q,\omega}[y](t) + \varepsilon D_{q,\omega}[h](t))d_{q,\omega}t$$

and

$$\mathcal{L}_a[y + \varepsilon h] = \int_{\omega_0}^a L(t, y(qt + \omega) + \varepsilon h(qt + \omega), D_{q,\omega}[y](t) + \varepsilon D_{q,\omega}[h](t))d_{q,\omega}t,$$

we have

$$\mathcal{L}[y + \varepsilon h] = \mathcal{L}_b[y + \varepsilon h] - \mathcal{L}_a[y + \varepsilon h].$$

Therefore,

$$\delta\mathcal{L}[y, h] = \delta\mathcal{L}_b[y, h] - \delta\mathcal{L}_a[y, h]. \tag{2.3}$$

In order to simplify expressions, we introduce the operator $\{\cdot\}$ defined in the following way:

$$\{y\}(t) := (t, y(qt + \omega), D_{q,\omega}[y](t)),$$

where $y \in \mathcal{Y}^1$.

Knowing (2.3), the following lemma is a direct consequence of Lemma 2.15.

Lemma 2.19 *For fixed* $y, h \in \mathcal{Y}^1$ *let*

$$g(t, \varepsilon) = L(t, y(qt + \omega) + \varepsilon h(qt + \omega), D_{q,\omega}[y](t) + \varepsilon D_{q,\omega}[h](t))$$

for $\varepsilon \in] -\bar{\varepsilon}, \bar{\varepsilon}[$, *for some* $\bar{\varepsilon} > 0$, *i.e.,*

$$g(t, \varepsilon) = L\{y + \varepsilon h\}(t).$$

Assume that:

(i) $g(t, \cdot)$ *is differentiable at* 0 *uniformly in* $t \in [a, b]_{q,\omega}$;

(ii) $\mathcal{L}_a[y + \varepsilon h] = \displaystyle\int_{\omega_0}^{a} g(t, \epsilon) \, d_{q,\omega} t$ *and* $\mathcal{L}_b[y + \varepsilon h] = \displaystyle\int_{\omega_0}^{b} g(t, \epsilon) \, d_{q,\omega} t$ *exist for* $\varepsilon \approx 0$;

(iii) $\displaystyle\int_{\omega_0}^{a} \partial_2 g(t, 0) d_{q,\omega} t$ *and* $\displaystyle\int_{\omega_0}^{b} \partial_2 g(t, 0) d_{q,\omega} t$ *exist.*

Then,

$$\delta\mathcal{L}[y, h] = \int_{a}^{b} \Big(\partial_2 L\{y\}(t) \cdot h(qt + \omega) + \partial_3 L\{y\}(t) \cdot D_{q,\omega}[h](t) \Big) d_{q,\omega} t.$$

The following result offers a necessary condition for local extremizer.

Theorem 2.20 (A necessary optimality condition for problem (2.1)–(2.2)**)** *Suppose that the optimal path to problem* (2.1)–(2.2) *exists and is given by* $\tilde{y}$. *Then,* $\delta\mathcal{L}[\tilde{y}, h] = 0$.

Proof Without loss of generality, we can assume $\tilde{y}$ to be a local minimizer. Let h be any admissible variation and define a function $\phi :]-\bar{\varepsilon}, \bar{\varepsilon}[\rightarrow \mathbb{R}$ by $\phi(\varepsilon) = \mathcal{L}[\tilde{y} + \varepsilon h]$. Since $\tilde{y}$ is a local minimizer, there exists $\delta > 0$, such that $\mathcal{L}[\tilde{y}] \leq \mathcal{L}[y]$ for all admissible y with $\|y - \tilde{y}\|_{1,\infty} < \delta$. Therefore, $\phi(\varepsilon) = \mathcal{L}[\tilde{y} + \varepsilon h] \geq \mathcal{L}[\tilde{y}] = \phi(0)$ for all $\varepsilon < \frac{\delta}{\|h\|_{1,\infty}}$. Hence, ϕ has a local minimum at $\varepsilon = 0$, and thus our assertion follows.

Theorem 2.21 (The Hahn quantum Euler–Lagrange equation for problem (2.1)–(2.2)**)** *Under hypotheses* (H1)–(H3) *and conditions* (i)–(iii) *of Lemma* 2.19 *on the Lagrangian* L, *if* $\tilde{y}$ *is a local minimizer or local maximizer to problem* (2.1)–(2.2), *then* $\tilde{y}$ *satisfies the Euler–Lagrange equation*

$$\partial_2 L\{y\}(t) - D_{q,\omega}[\partial_3 L]\{y\}(t) = 0 \tag{2.4}$$

for all $t \in [a, b]_{q,\omega}$.

Proof Suppose that $\mathcal{L}$ has a local extremum at $\tilde{y}$. Let h be any admissible variation and define a function $\phi :]-\bar{\varepsilon}, \bar{\varepsilon}[\rightarrow \mathbb{R}$ by $\phi(\varepsilon) = \mathcal{L}[\tilde{y} + \varepsilon h]$. A necessary condition for $\tilde{y}$ to be an extremizer is given by $\phi'(0) = 0$. Note that

$$\phi'(0) = \int_a^b \Big(\partial_2 L\{\tilde{y}\}(t) \cdot h(qt + \omega) + \partial_3 L\{\tilde{y}\}(t) \cdot D_{q,\omega}[h](t) \Big) d_{q,\omega} t.$$

Since $h(a) = h(b) = 0$, then

$$\phi'(0) = \int_a^b \Big(\partial_2 L\{\tilde{y}\}(t) \cdot h(qt + \omega) + \partial_3 L\{\tilde{y}\}(t) \cdot D_{q,\omega}[h](t) \Big) d_{q,\omega} t.$$

Integration by parts gives

$$\int_a^b \partial_3 L\{\tilde{y}\}(t) \cdot D_{q,\omega}[h](t) d_{q,\omega} t = \Big[\partial_3 L\{\tilde{y}\}(t) \cdot h(t) \Big]_a^b$$
$$- \int_a^b D_{q,\omega}[\partial_3 L]\{\tilde{y}\}(t) \cdot h(qt + \omega) d_{q,\omega} t$$

and since $h(a) = h(b) = 0$, then

$$\phi'(0) = 0 \Leftrightarrow \int_a^b \Big(\partial_2 L\{\tilde{y}\}(t) - D_{q,\omega}[\partial_3 L]\{\tilde{y}\}(t) \Big) \cdot h(qt + \omega) d_{q,\omega} t = 0.$$

Thus, by Lemma 2.16, we have

$$\partial_2 L\{\tilde{y}\}(t) - D_{q,\omega}[\partial_3 L]\{\tilde{y}\}(t) = 0$$

for all $t \in [a, b]_{q,\omega}$.

Remark 2.22 Under appropriate conditions, when $(\omega, q) \rightarrow (0, 1)$, we obtain a corresponding result in the classical context of the calculus of variations (1.4):

$$\frac{d}{dt} \partial_3 L(t, y(t), y'(t)) = \partial_2 L(t, y(t), y'(t)).$$

Remark 2.23 In practical terms the hypotheses of Theorem 2.21 are not easy to verify *a priori*. However, we can assume that all hypotheses are satisfied and apply the q, ω-Euler–Lagrange equation (2.4) heuristically to obtain a *candidate*. If such a candidate is, or not, a solution to the variational problem is a different question that require further analysis (see Sects. 2.4 and 2.8.5).

2.3 The Hahn Quantum Isoperimetric Problem

We now study the isoperimetric problem with an integral constraint. Both normal and abnormal extremizers are considered. Isoperimetric problems have found a broad class of important applications throughout the centuries. Areas of application include also economy (see, e.g., Almeida and Torres (2009b); Caputo (2005) and the references given there).

The isoperimetric problem consists of minimizing or maximizing the functional (2.1) in the class of functions $y \in \mathcal{Y}^1$ satisfying the boundary conditions (2.2), and the integral constraint

$$\mathcal{J}[y] = \int_a^b F\left(t, y\left(qt + \omega\right), D_{q,\omega}[y](t)\right) d_{q,\omega}t = \gamma \qquad (2.5)$$

for some $\gamma \in \mathbb{R}$.

Definition 2.24 We say that $\tilde{y} \in \mathcal{Y}^1$ is a local minimizer (resp. local maximizer) for the isoperimetric problem (2.1), (2.2) and (2.5) if there exists $\delta > 0$ such that $\mathcal{L}[\tilde{y}] \leq \mathcal{L}[y]$ (resp. $\mathcal{L}[\tilde{y}] \geq \mathcal{L}[y]$) for all $y \in \mathcal{Y}^1$ satisfying the boundary conditions (2.2) and the isoperimetric constraint (2.5) and $\|\tilde{y} - y\|_{1,\infty} < \delta$.

Definition 2.25 We say that $y \in \mathcal{Y}^1$ is an extremal to $\mathcal{J}$ if y satisfies the Euler–Lagrange equation (2.4) relatively to $\mathcal{J}$. An extremizer (i.e., a local minimizer or a local maximizer) to problem (2.1), (2.2) and (2.5) that is not an extremal to $\mathcal{J}$ is said to be a normal extremizer; otherwise, the extremizer is said to be abnormal.

Theorem 2.26 (Necessary optimality condition for normal extremizers to (2.1)**,** (2.2) **and** (2.5)**)** *Suppose that L and F satisfy hypotheses (H1)–(H3) and conditions (i)–(iii) of Lemma 2.19, and suppose that $\tilde{y} \in \mathcal{Y}^1$ gives a local minimum or a local maximum to the functional $\mathcal{L}$ subject to the integral constraint (2.5). If $\tilde{y}$ is not an extremal to $\mathcal{J}$, then there exists a real number λ such that $\tilde{y}$ satisfies the equation*

$$\partial_2 H\{y\}(t) - D_{q,\omega}[\partial_3 H]\{y\}(t) = 0 \qquad (2.6)$$

for all $t \in [a, b]_{q,\omega}$, where $H = L - \lambda F$.

Proof Suppose that $\tilde{y} \in \mathcal{Y}^1$ is a normal extremizer to problem (2.1), (2.2) and (2.5). Define the real functions $\phi, \psi : \mathbb{R}^2 \to \mathbb{R}$ by

$$\phi(\epsilon_1, \epsilon_2) = \mathcal{L}[\tilde{y} + \epsilon_1 h_1 + \epsilon_2 h_2],$$
$$\psi(\epsilon_1, \epsilon_2) = \mathcal{J}[\tilde{y} + \epsilon_1 h_1 + \epsilon_2 h_2] - \gamma,$$

where $h_2 \in \mathcal{Y}^1$ is fixed (that we will choose later) and $h_1 \in \mathcal{Y}^1$ is an arbitrary function. Note that

$$\frac{\partial \psi}{\partial \epsilon_2}(0,0) = \int_a^b \left(\partial_2 F\{\widetilde{y}\}(t) \cdot h_2(qt + \omega) + \partial_3 F\{\widetilde{y}\}(t) \cdot D_{q,\omega}[h_2](t) \right) d_{q,\omega}t.$$

Using integration by parts formula we get

$$\frac{\partial \psi}{\partial \epsilon_2}(0,0) = \int_a^b \left(\partial_2 F\{\widetilde{y}\}(t) - D_{q,\omega}[\partial_3 F]\{\widetilde{y}\}(t) \right) \cdot h_2(qt + \omega) d_{q,\omega}t$$
$$+ \left[\partial_3 F\{\widetilde{y}\}(t) \cdot h_2(t) \right]_a^b.$$

Restricting h_2 to those such that $h_2(a) = h_2(b) = 0$ we obtain

$$\frac{\partial \psi}{\partial \epsilon_2}(0,0) = \int_a^b \left(\partial_2 F\{\widetilde{y}\}(t) - D_{q,\omega}[\partial_3 F]\{\widetilde{y}\}(t) \right) \cdot h_2(qt + \omega) d_{q,\omega}t.$$

Since $\widetilde{y}$ is not an extremal to $\mathcal{J}$, then we can choose h_2 such that $\dfrac{\partial \psi}{\partial \epsilon_2}(0,0) \neq 0$.
We keep h_2 fixed. Since $\psi(0,0) = 0$, by the Implicit Function Theorem there exists a function g defined in a neighborhood V of zero, such that $g(0) = 0$ and $\psi(\epsilon_1, g(\epsilon_1)) = 0$, for any $\epsilon_1 \in V$, that is, there exists a subset of variation curves $y = \widetilde{y} + \epsilon_1 h_1 + g(\epsilon_1)h_2$ satisfying the isoperimetric constraint. Note that $(0,0)$ is an extremizer of ϕ subject to the constraint $\psi = 0$ and

$$\nabla \psi(0,0) \neq (0,0).$$

By the Lagrange multiplier rule, there exists some constant $\lambda \in \mathbb{R}$ such that

$$\nabla \phi(0,0) = \lambda \nabla \psi(0,0). \tag{2.7}$$

Restricting h_1 to those such that $h_1(a) = h_1(b) = 0$ we get

$$\frac{\partial \phi}{\partial \epsilon_1}(0,0) = \int_a^b \left(\partial_2 L\{\widetilde{y}\}(t) - D_{q,\omega}[\partial_3 L]\{\widetilde{y}\}(t) \right) \cdot h_1(qt + \omega) d_{q,\omega}t$$

and

$$\frac{\partial \psi}{\partial \epsilon_1}(0,0) = \int_a^b \left(\partial_2 F\{\widetilde{y}\}(t) - D_{q,\omega}[\partial_3 F]\{\widetilde{y}\}(t) \right) \cdot h_1(qt + \omega) d_{q,\omega}t.$$

Using (2.7) it follows that

$$\int_a^b \Big(\partial_2 L\{\widetilde{y}\}(t) - D_{q,\omega}[\partial_3 L]\{\widetilde{y}\}(t)$$
$$- \lambda \Big(\partial_2 F\{\widetilde{y}\}(t) - D_{q,\omega}[\partial_3 F]\{\widetilde{y}\}(t) \Big) \Big) \cdot h_1(qt + \omega) d_{q,\omega}t = 0.$$

Using the fundamental lemma of the Hahn quantum variational calculus (Lemma 2.16), and recalling that h_1 is arbitrary, we conclude that

$$\partial_2 L\{\widetilde{y}\}(t) - D_{q,\omega}[\partial_3 L]\{\widetilde{y}\}(t) - \lambda\left(\partial_2 F\{\widetilde{y}\}(t) - D_{q,\omega}[\partial_3 F]\{\widetilde{y}\}(t)\right) = 0$$

for all $t \in [a, b]_{q,\omega}$, proving that $H = L - \lambda F$ satisfies the Euler–Lagrange condition (2.6).

Introducing an extra multiplier λ_0 we can also deal with abnormal extremizers to the isoperimetric problem (2.1), (2.2) and (2.5).

Theorem 2.27 (Necessary optimality condition for normal and abnormal extremizers to (2.1), (2.2) and (2.5)) *Suppose that L and F satisfy hypotheses (H1)–(H3) and conditions (i)–(iii) of Lemma 2.19, and suppose that $\widetilde{y} \in \mathcal{Y}^1$ gives a local minimum or a local maximum to the functional $\mathcal{L}$ subject to the integral constraint (2.5). Then there exist two constants λ_0 and λ, not both zero, such that $\widetilde{y}$ satisfies the equation*

$$\partial_2 H\{y\}(t) - D_{q,\omega}[\partial_3 H]\{y\}(t) = 0 \tag{2.8}$$

for all $t \in [a, b]_{q,\omega}$, where $H = \lambda_0 L - \lambda F$.

Proof The proof is similar to the proof of Theorem 2.26. Since $(0, 0)$ is an extremizer of ϕ subject to the constraint $\psi = 0$, the abnormal Lagrange multiplier rule (cf., e.g., van Brunt (2004)) guarantees the existence of two reals λ_0 and λ, not both zero, such that

$$\lambda_0 \nabla\phi = \lambda \nabla\psi.$$

Remark 2.28 Note that if $\widetilde{y}$ is a normal extremizer then, by Theorem 2.26, one can choose $\lambda_0 = 1$ in Theorem 2.27. The condition $(\lambda_0, \lambda) \neq (0, 0)$ guarantees that Theorem 2.27 is a useful necessary condition. In general we cannot guarantee, a priori, that λ_0 be different from zero. The interested reader about abnormality is referred to the book (Arutyunov 2000).

Suppose now that it is required to find functions y_1 and y_2 for which the functional

$$\mathcal{L}[y_1, y_2] = \int_a^b f(t, y_1(qt+\omega), y_2(qt+\omega), D_{q,\omega}[y_1](t), D_{q,\omega}[y_2](t))d_{q,\omega}t \tag{2.9}$$

has an extremum, where the admissible functions satisfy the boundary conditions

$$(y_1(a), y_2(a)) = (y_1^a, y_2^a) \text{ and } (y_1(b), y_2(b)) = (y_1^b, y_2^b), \tag{2.10}$$

and the subsidiary nonholonomic condition

$$g(t, y_1(qt+\omega), y_2(qt+\omega), D_{q,\omega}[y_1](t), D_{q,\omega}[y_2](t)) = 0. \tag{2.11}$$

The problem (2.9)–(2.11) can be reduced to the isoperimetric one by transforming (2.11) into a constraint of the type (2.5). For that, we multiply both sides of (2.11) by an arbitrary function $\lambda(t)$, and then take the q, ω-integral from a to b. We obtain the new constraint

$$\mathcal{K}[y_1, y_2] = \int_a^b \lambda(t) g(t, y_1(qt + \omega), y_2(qt + \omega), D_{q,\omega}[y_1](t), D_{q,\omega}[y_2](t)) d_{q,\omega} t = 0.$$

$$(2.12)$$

Under the conditions of Theorem 2.26, the solutions (y_1, y_2) of the isoperimetric problem (2.9) and (2.12) satisfy the Euler–Lagrange equation for the functional

$$\int_a^b (f - \tilde{\lambda}(t)g) d_{q,\omega} t,$$

$$(2.13)$$

$\tilde{\lambda}(t) = \bar{\lambda}\lambda(t)$ for some constant $\bar{\lambda}$. Since (2.12) follows from (2.11), the solutions of problem (2.9)–(2.11) satisfy the Euler–Lagrange equation for functional (2.13) as well.

2.4 Sufficient Condition for Optimality

In this subsection we prove sufficient optimality conditions for problem (2.1)–(2.2). Similar to the classical calculus of variations we assume the Lagrangian function to be convex (or concave).

Theorem 2.29 *Let* $L(t, u_0, u_1)$ *be jointly convex (resp. concave) in* (u_0, u_1). *If* $\tilde{y}$ *satisfies condition (2.4), then* $\tilde{y}$ *is a global minimizer (resp. maximizer) to problem (2.1)–(2.2).*

Proof We give the proof for the convex case. Since L is jointly convex in (u_0, u_1), then for any $h \in \mathcal{Y}^1$,

$$\mathcal{L}[\tilde{y} + h] - \mathcal{L}[\tilde{y}] = \int_a^b \left(L\{\tilde{y} + h\}(t) - L\{\tilde{y}\}(t) \right) d_{q,\omega} t$$

$$\geq \int_a^b \left(\partial_2 L\{\tilde{y}\}(t) \cdot h(qt + \omega) + \partial_3 L\{\tilde{y}\}(t) \cdot D_{q,\omega}[h](t) \right) d_{q,\omega} t.$$

Proceeding analogously as in the proof of Theorem 2.21 and since $\tilde{y}$ satisfies condition (2.4) we obtain $\mathcal{L}(\tilde{y} + h) - \mathcal{L}(\tilde{y}) \geq 0$, proving the desired result.

2.5 Leitmann's Direct Method

Leitmann's direct method permits to compute global solutions to some problems that are variationally invariant under a family of transformations (Leitmann 1967, 2001a,b; Silva and Torres 2006; Torres and Leitmann 2008; Wagener 2009). It should be mentioned that such invariance transformations are useful not only in connection with Leitmann's method but also to apply Noether's Theorem (Torres 2002, 2004a). Moreover, the invariance transformations are related with the notion of Carathéodory equivalence (Carlson 2002; Torres 2004b).

Recently, it has been noticed in Malinowska and Torres (2010a) that the invariance transformations, that keep the Lagrangian invariant, do not depend on the time scale. This is also true for the generalized Hahn quantum setting that we are considering in this work: given a Lagrangian $L : \mathbb{R} \times \mathbb{R} \times \mathbb{R} \to \mathbb{R}$, the invariance transformations, that keep it invariant up to a gauge term, are exactly the same if the Lagrangian L is used to define a Hahn quantum functional (2.1) or a classical functional $\mathcal{L}[y] = \int_a^b L(t, y(t), y'(t))dt$ of the calculus of variations. Thus, if the quantum problem we want to solve admits an enough rich family of invariance transformations, that keep it invariant up to a gauge term, then one does not need to solve a Hahn quantum Euler–Lagrange equation to find its minimizer: instead, we can try to use Leitmann's direct method. The question of how to find the invariance transformations is addressed in Gouveia and Torres (2005); Gouveia et al. (2006).

Let $\bar{L} : [a, b] \times \mathbb{R} \times \mathbb{R} \to \mathbb{R}$. We assume that $\bar{L}$ satisfies hypotheses (H1)–(H3). Consider the integral

$$\bar{\mathcal{L}}[\bar{y}] = \int_a^b \bar{L}\{\bar{y}\}(t)d_{q,\omega}t.$$

Lemma 2.30 (Leitmann's fundamental lemma via Hahn's quantum operator)
Let $y = z(t, \bar{y})$ be a transformation having an unique inverse $\bar{y} = \bar{z}(t, y)$ for all $t \in [a, b]$, such that there is a one-to-one correspondence

$$y(t) \Leftrightarrow \bar{y}(t)$$

for all functions $y \in \mathcal{Y}^1$ satisfying (2.2) and all functions $\bar{y} \in \mathcal{Y}^1$ satisfying

$$\bar{y}(a) = \bar{z}(a, \alpha), \quad \bar{y}(b) = \bar{z}(b, \beta). \tag{2.14}$$

If the transformation $y = z(t, \bar{y})$ is such that there exists a function $G : [a, b] \times \mathbb{R} \to \mathbb{R}$ satisfying the functional identity

$$L\{y\}(t) - \bar{L}\{\bar{y}\}(t) = D_{q,\omega}G(t, \bar{y}(t)), \tag{2.15}$$

then if $\bar{y}^*$ yields the extremum of $\bar{\mathcal{L}}$ with $\bar{y}^*$ satisfying (2.14), $y^* = z(t, \bar{y}^*)$ yields the extremum of $\mathcal{L}$ for y^* satisfying (2.2).

Remark 2.31 The functional identity (2.15) is exactly the definition of variationally invariance when we do not consider transformations of the time variable t (cf. (4) and (5) of Torres and Leitmann (2008)). Function G that appears in (2.15) is sometimes called a gauge term (Torres 2004a).

Proof The proof is similar in spirit to Leitmann's proof (Leitmann 1967, 2001a,b, 2004). Let $y \in \mathcal{Y}^1$ satisfy (2.2), and define functions $\bar{y} \in \mathcal{Y}^1$ through the formula $\bar{y} = \bar{z}(t, y), t \in [a, b]$. Then $\bar{y} \in \mathcal{Y}^1$ and satisfies (2.14). Moreover, as a result of (2.15), it follows that

$$\mathcal{L}[y] - \bar{\mathcal{L}}[\bar{y}] = \int_a^b L\{y\}(t)d_{q,\omega}t - \int_a^b \bar{L}\{\bar{y}\}(t)d_{q,\omega}t = \int_a^b D_{q,\omega}G(t, \bar{y}(t))d_{q,\omega}t$$
$$= G(b, \bar{y}(b)) - G(a, \bar{y}(a)) = G(b, \bar{z}(b, \beta)) - G(a, \bar{z}(a, \alpha)),$$

from which the desired conclusion follows immediately since the right-hand side of the above equality is a constant, depending only on the fixed-endpoint conditions (2.2).

Examples 2.33, 2.34 and 2.35 in the next section illustrate the applicability of Lemma 2.30. The procedure is as follows: (i) we use the computer algebra package described in Gouveia and Torres (2005) and available from the *Maple Application Center* at http://www.maplesoft.com/applications/view.aspx?SID=4805 to find the transformations that keep the problem of the calculus of variations or optimal control invariant; (ii) we use such invariance transformations to solve the Hahn quantum variational problem by applying Leitmann's fundamental lemma (Lemma 2.30).

2.6 Illustrative Examples

We provide some examples in order to illustrate our main results.

Example 2.32 Let q, ω be fixed real numbers, and I be a closed interval of $\mathbb{R}$ such that $\omega_0, 0, 1 \in I$. Consider the problem

$$\mathcal{L}[y] = \int_0^1 \left(y(qt + \omega) + \frac{1}{2}(D_{q,\omega}[y](t))^2 \right) d_{q,\omega}t \longrightarrow \min \qquad (2.16)$$

subject to the boundary conditions

$$y(0) = 0, \quad y(1) = 1. \qquad (2.17)$$

If y is a local minimizer to problem (2.16)–(2.17), then by Theorem 2.21 it satisfies the Euler–Lagrange equation

$$D_{q,\omega}D_{q,\omega}[y](t) = 1 \qquad (2.18)$$

for all $t \in \{w[n]_q : n \in \mathbb{N}_0\} \cup \{q^n + w[n]_q : n \in \mathbb{N}_0\} \cup \{w_0\}$. By direct substitution it can be verified that $y(t) = \frac{1}{q+1}t^2 + \frac{q}{q+1}t$ is a candidate solution to problem (2.16)–(2.17).

In next examples we solve quantum variational problems using Leitmann's direct method (see Sect. 2.5).

Example 2.33 Let q, w, and a, b $(a < b)$ be fixed real numbers, and I be a closed interval of $\mathbb{R}$ such that $w_0 \in I$ and $a, b \in \{q^n s + w[n]_q : n \in \mathbb{N}_0\} \cup \{w_0\}$ for some $s \in I$. Let α and β be two given real numbers, $\alpha \neq \beta$. We consider the following problem:

$$\mathcal{L}[y] = \int_a^b \left((D_{q,w}[y](t))^2 + y(qt + w) + t D_{q,w}[y](t) \right) d_{q,w}t \longrightarrow \min \tag{2.19}$$
$$y(a) = \alpha, \quad y(b) = \beta.$$

We transform problem (2.19) into the trivial problem

$$\bar{\mathcal{L}}[\bar{y}] = \int_a^b (D_{q,w}[\bar{y}](t))^2 d_{q,w}t \longrightarrow \min$$
$$\bar{y}(a) = 0, \quad \bar{y}(b) = 0,$$

which has solution $\bar{y} \equiv 0$. For that we consider the transformation

$$y(t) = \bar{y}(t) + ct + d, \quad c, d \in \mathbb{R},$$

where constants c and d will be chosen later. According to the above, we have

$$D_{q,w}[y](t) = D_{q,w}[\bar{y}](t) + c, \quad y(qt + w) = \bar{y}(qt + w) + c(qt + w) + d,$$

and

$$(D_{q,w}[y](t))^2 + y(qt + w) + t D_{q,w}[y](t)$$
$$= (D_{q,w}[\bar{y}](t))^2 + 2c D_{q,w}[\bar{y}](t) + c^2 + \bar{y}(qt + w) + c(qt + w) + d$$
$$+ t D_{q,w}[\bar{y}](t) + ct$$
$$= (D_{q,w}[\bar{y}](t))^2 + D_{q,w}[2c\bar{y}(t) + t\bar{y}(t) + ct^2 + (c^2 + d)t].$$

In order to obtain the solution to the original problem, it suffices to chose c and d so that

$$ca + d = \alpha, \quad cb + d = \beta. \tag{2.20}$$

Solving the system of equations (2.20) we obtain $c = \frac{\alpha - \beta}{a - b}$ and $d = \frac{\beta a - b\alpha}{a - b}$. Hence, the global minimizer for problem (2.19) is

$$y(t) = \frac{\alpha - \beta}{a - b} t + \frac{\beta a - b\alpha}{a - b}.$$

Example 2.34 Let q, ω, and a, b ($a < b$) be fixed real numbers, and I be a closed interval of $\mathbb{R}$ such that $\omega_0 \in I$ and $a, b \in \{q^n s + \omega[n]_q : n \in \mathbb{N}_0\} \cup \{\omega_0\}$ for some $s \in I$. Let α and β be two given real numbers, $\alpha \neq \beta$. We consider the following problem:

$$\mathcal{L}[y] = \int_a^b \left(D_{q,\omega}[yg](t) \right)^2 d_{q,\omega}t \longrightarrow \min \tag{2.21}$$
$$y(a) = \alpha, \quad y(b) = \beta,$$

where g does not vanish on the interval $[a, b]_{q,\omega}$. Observe that $\bar{y}(t) = g^{-1}(t)$ minimizes $\mathcal{L}$ with end conditions $\bar{y}(a) = g^{-1}(a)$ and $\bar{y}(b) = g^{-1}(b)$. Let $y(t) = \bar{y}(t) + p(t)$. Then

$$\left(D_{q,\omega}[yg](t) \right)^2 = \left(D_{q,\omega}[\bar{y}g](t) \right)^2 + D_{q,\omega}[pg](t) D_{q,\omega}[2\bar{y}g + pg](t). \tag{2.22}$$

Consequently, if $p(t) = (At + B)g^{-1}(t)$, where A and B are constants, then (2.22) is of the form (2.15), since $D_{q,\omega}[pg](t)$ is constant. Thus, the function

$$y(t) = (At + C)g^{-1}(t)$$

with

$$A = [\alpha g(a) - \beta g(b)] (a - b)^{-1}, \quad C = [a\beta g(b) - b\alpha g(a)] (a - b)^{-1},$$

minimizes (2.21).

Using the idea of Leitmann, we can also solve quantum optimal control problems defined in terms of Hahn's operators.

Example 2.35 Let q, ω be real numbers on a closed interval I of $\mathbb{R}$ such that $\omega_0 \in I$ and $0, 1 \in \{q^n s + \omega[n]_q : n \in \mathbb{N}_0\} \cup \{\omega_0\}$ for some $s \in I$. Consider the global minimum problem

$$\mathcal{L}[u_1, u_2] = \int_0^1 \left((u_1(t))^2 + u_2(t))^2 \right) d_{q,\omega}t \longrightarrow \min \tag{2.23}$$

subject to the control system

$$D_{q,\omega}[y_1](t) = \exp(u_1(t)) + u_1(t) + u_2(t), \quad D_{q,\omega}[y_2](t) = u_2(t), \tag{2.24}$$

and conditions

$$y_1(0) = 0, \quad y_1(1) = 2, \quad y_2(0) = 0, \quad y_2(1) = 1, \tag{2.25}$$
$$u_1(t), u_2(t) \in \Omega = [-1, 1].$$

This example is inspired from Torres and Leitmann (2008). It is worth mentioning that due to the constraints on the values of the controls, $(u_1(t), u_2(t)) \in \Omega = [-1, 1]$, a theory based on necessary optimality conditions to solve problem (2.23)–(2.25) does not exist at the moment.

We begin noticing that problem (2.23)–(2.25) is variationally invariant according to Gouveia and Torres (2005) under the one-parameter family of transformations

$$y_1^s = y_1 + st, \quad y_2^s = y_2 + st, \quad u_2^s = u_2 + s \quad (t^s = t \text{ and } u_1^s = u_1). \qquad (2.26)$$

To prove this, we need to show that both the functional integral $\mathcal{L}$ and the control system stay invariant under the s-parameter transformations (2.26). This is easily seen by direct calculations:

$$
\begin{aligned}
\mathcal{L}^s[u_1^s, u_2^s] &= \int_0^1 \left(u_1^s(t)\right)^2 + \left(u_2^s(t)\right)^2 d_{q,\omega}t \\
&= \int_0^1 u_1^2(t) + (u_2(t) + s)^2 d_{q,\omega}t \qquad (2.27) \\
&= \int_0^1 \left(u_1^2(t) + u_2^2(t) + D_{q,\omega}[s^2t + 2sy_2(t)]\right) d_{q,\omega}t \\
&= \mathcal{L}[u_1, u_2] + s^2 + 2s.
\end{aligned}
$$

We remark that $\mathcal{L}^s$ and $\mathcal{L}$ have the same minimizers: adding a constant $s^2 + 2s$ to the functional $\mathcal{L}$ does not change the minimizer of $\mathcal{L}$. It remains to prove that the control system also remains invariant under transformations (2.26):

$$
\begin{aligned}
D_{q,\omega}[y_1^s](t) &= D_{q,\omega}[y_1](t) + s = \exp(u_1(t)) + u_1(t) + u_2(t) + s \\
&= \exp(u_1^s(t)) + u_1^s(t) + u_2^s(t), \qquad (2.28) \\
D_{q,\omega}[y_2^s](t) &= D_{q,\omega}[y_2](t) + s = u_2(t) + s = u_2^s(t).
\end{aligned}
$$

Conditions (2.27) and (2.28) prove that problem (2.23)–(2.25) is invariant under the s-parameter transformations (2.26) up to $D_{q,\omega}\left(s^2t + 2sy_2(t)\right)$. Using the invariance transformations (2.26), we generalize problem (2.23)–(2.25) to a s-parameter family of problems, $s \in \mathbb{R}$, which include the original problem for $s = 0$:

$$\mathcal{L}^s[u_1, u_2] = \int_0^1 (u_1^s(t))^2 + (u_2^s(t))^2 d_{q,\omega}t \longrightarrow \min$$

subject to the control system

$$D_{q,\omega}[y_1^s](t) = \exp(u_1^s(t)) + u_1^s(t) + u_2^s(t), \quad D_{q,\omega}[y_2^s(t)] = u_2^s(t),$$

and conditions

$$y_1^s(0) = 0, \quad y_1^s(1) = 2 + s, \quad y_2^s(0) = 0, \quad y_2^s(1) = 1 + s,$$
$$u_1^s(t) \in [-1, 1], \quad u_2^s(t) \in [-1 + s, 1 + s].$$

It is clear that $\mathcal{L}^s \geq 0$ and that $\mathcal{L}^s = 0$ if $u_1^s(t) = u_2^s(t) \equiv 0$. The control equations, the boundary conditions and the constraints on the values of the controls imply that $u_1^s(t) = u_2^s(t) \equiv 0$ is admissible only if $s = -1$: $y_1^{s=-1}(t) = t, y_2^{s=-1}(t) \equiv 0$. Hence, for $s = -1$ the global minimum to $\mathcal{L}^s$ is 0 and the minimizing trajectory is given by

$$\tilde{u}_1^s(t) \equiv 0, \quad \tilde{u}_2^s(t) \equiv 0, \quad \tilde{y}_1^s(t) = t, \quad \tilde{y}_2^s(t) \equiv 0.$$

Since for any s one has by (2.27) that $\mathcal{L}[u_1, u_2] = \mathcal{L}^s[u_1^s, u_2^s] - s^2 - 2s$, we conclude that the global minimum for problem $\mathcal{L}[u_1, u_2]$ is 1. Thus, using the inverse functions of the variational symmetries (2.26),

$$u_1(t) = u_1^s(t), \quad u_2(t) = u_2^s(t) - s, \quad y_1(t) = y_1^s(t) - st, \quad y_2(t) = y_2^s(t) - st,$$

and the absolute minimizer for problem (2.23)–(2.25) is

$$\tilde{u}_1(t) = 0, \quad \tilde{u}_2(t) = 1, \quad \tilde{y}_1(t) = 2t, \quad \tilde{y}_2(t) = t.$$

2.7 Higher-order Hahn's Quantum Variational Calculus

We define the q, ω-derivatives of higher-order in the usual way: the rth q, ω-derivative ($r \in \mathbb{N}$) of $f : I \to \mathbb{R}$ is the function $D_{q,\omega}^r[f] : I \to \mathbb{R}$ given by $D_{q,\omega}^r[f] := D_{q,\omega}[D_{q,\omega}^{r-1}[f]]$, provided $D_{q,\omega}^{r-1}[f]$ is q, ω-differentiable on I and where $D_{q,\omega}^0[f] := f$. The following notations are in order: $\sigma(t) = qt + \omega$, $y^{\sigma}(t) = y^{\sigma^1}(t) = (y \circ \sigma)(t) = y(qt + \omega)$, and $y^{\sigma^k} = y \circ y^{\sigma^{k-1}}, k = 2, 3, \ldots$

Our main goal is to establish necessary optimality conditions for the higher-order q, ω-variational problem

$$\mathcal{L}[y] = \int_a^b L\left(t, y^{\sigma^r}(t), D_{q,\omega}\left[y^{\sigma^{r-1}}\right](t), \ldots, D_{q,\omega}^r[y](t)\right) d_{q,\omega}t \longrightarrow \text{extr}$$

$$y \in \mathcal{Y}^r([a, b], \mathbb{R}) \tag{P}$$

$$y(a) = \alpha_0, \quad y(b) = \beta_0,$$

$$\vdots$$

$$D_{q,\omega}^{r-1}[y](a) = \alpha_{r-1}, \quad D_{q,\omega}^{r-1}[y](b) = \beta_{r-1},$$

where $r \in \mathbb{N}$ and $\alpha_i, \beta_i \in \mathbb{R}, i = 0, \ldots, r - 1$, are given.

Definition 2.36 We say that y is an admissible function for (2.7) if $y \in \mathcal{Y}^r([a, b], \mathbb{R})$ and y satisfies the boundary conditions $D_{q,\omega}^i[y](a) = \alpha_i$ and $D_{q,\omega}^i[y](b) = \beta_i$ of problem (2.7), $i = 0, \ldots, r - 1$.

The Lagrangian L is assumed to satisfy the following hypotheses:

(H1) $(u_0, \ldots, u_r) \to L(t, u_0, \ldots, u_r)$ is a $C^1(\mathbb{R}^{r+1}, \mathbb{R})$ function for any $t \in [a, b]$;

(H2) $t \to L(t, y(t), D_{q,\omega}[y](t), \ldots, D_{q,\omega}^r[y](t))$ is continuous at ω_0 for any admissible y;

(H3) functions $t \to \partial_{i+2} L(t, y(t), D_{q,\omega}[y](t), \cdots, D_{q,\omega}^r[y](t)), i = 0, 1, \cdots, r$, belong to $\mathcal{Y}^1([a, b], \mathbb{R})$ for all admissible y.

Definition 2.37 We say that y_* is a local minimizer (resp. local maximizer) for problem (2.7) if y_* is an admissible function and there exists $\delta > 0$ such that

$$\mathcal{L}[y_*] \le \mathcal{L}[y] \quad (\text{resp. } \mathcal{L}[y_*] \ge \mathcal{L}[y])$$

for all admissible y with $\|y_* - y\|_{r,\infty} < \delta$.

Definition 2.38 We say that $\eta \in \mathcal{Y}^r([a, b], \mathbb{R})$ is a *variation* if $\eta(a) = \eta(b) = 0$, ..., $D_{q,\omega}^{r-1}[\eta](a) = D_{q,\omega}^{r-1}[\eta](b) = 0$.

2.7.1 Higher-order Fundamental Lemma

The chain rule, as known from classical calculus, does not hold in Hahn's quantum context (see a counterexample in Aldwoah (2009); Annaby et al. (2012)). However, we can prove the following.

Lemma 2.39 *If f is q, ω-differentiable on I, then the following equality holds:*

$$D_{q,\omega}[f^\sigma](t) = q(D_{q,\omega}[f])^\sigma(t), t \in I.$$

Proof For $t \ne \omega_0$ we have

$$(D_{q,\omega}[f])^\sigma(t) = \frac{f(q(qt + \omega) + \omega) - f(qt + \omega)}{(q - 1)(qt + \omega) + \omega}$$
$$= \frac{f(q(qt + \omega) + \omega) - f(qt + \omega)}{q((q - 1)t + \omega)}$$

and

$$D_{q,\omega}[f^\sigma](t) = \frac{f^\sigma(qt + \omega) - f^\sigma(t)}{(q - 1)t + \omega} = \frac{f(q(qt + \omega) + \omega) - f(qt + \omega)}{(q - 1)t + \omega}.$$

Therefore, $D_{q,\omega}[f^\sigma](t) = q(D_{q,\omega}[f])^\sigma(t)$. If $t = \omega_0$, then $\sigma(\omega_0) = \omega_0$. Thus,

$$\left(D_{q,\omega}\left[f\right]\right)^{\sigma}(\omega_0) = \left(D_{q,\omega}\left[f\right]\right)(\sigma(\omega_0)) = \left(D_{q,\omega}\left[f\right]\right)(\omega_0) = f'(\omega_0)$$

and $D_{q,\omega}\left[f^{\sigma}\right](\omega_0) = \left[f^{\sigma}\right]'(\omega_0) = f'(\sigma(\omega_0))\,\sigma'(\omega_0) = qf'(\omega_0)$.

Lemma 2.40 *If* $\eta \in \mathcal{Y}^r\left([a,b],\mathbb{R}\right)$ *is such that* $D_{q,\omega}^i\left[\eta\right](a) = 0$ *(resp.* $D_{q,\omega}^i\left[\eta\right]$ *(b) = 0) for all* $i \in \{0, 1, \ldots, r\}$, *then* $D_{q,\omega}^{i-1}\left[\eta^{\sigma}\right](a) = 0$ *(resp.* $D_{q,\omega}^{i-1}\left[\eta^{\sigma}\right](b) = 0$) *for all* $i \in \{1, \ldots, r\}$.

Proof If $a = \omega_0$ the result is trivial (because $\sigma(\omega_0) = \omega_0$). Suppose now that $a \neq \omega_0$ and fix $i \in \{1, \ldots, r\}$. Note that

$$D_{q,\omega}^i\left[\eta\right](a) = \frac{\left(D_{q,\omega}^{i-1}\left[\eta\right]\right)^{\sigma}(a) - D_{q,\omega}^{i-1}\left[\eta\right](a)}{(q-1)a + \omega}.$$

Because, by hypothesis, $D_{q,\omega}^i\left[\eta\right](a) = 0$ and $D_{q,\omega}^{i-1}\left[\eta\right](a) = 0$, then

$$\left(D_{q,\omega}^{i-1}\left[\eta\right]\right)^{\sigma}(a) = 0.$$

Lemma 2.39 shows that

$$\left(D_{q,\omega}^{i-1}\left[\eta\right]\right)^{\sigma}(a) = \left(\frac{1}{q}\right)^{i-1} D_{q,\omega}^{i-1}\left[\eta^{\sigma}\right](a).$$

We conclude that $D_{q,\omega}^{i-1}\left[\eta^{\sigma}\right](a) = 0$. The case $t = b$ is proved in the same way.

Lemma 2.41 *Suppose that* $f \in \mathcal{Y}^1\left([a,b],\mathbb{R}\right)$. *One has*

$$\int_a^b f(t)\,D_{q,\omega}\left[\eta\right](t)\,d_{q,\omega}t = 0$$

for all functions $\eta \in \mathcal{Y}^1\left([a,b],\mathbb{R}\right)$ *such that* $\eta(a) = \eta(b) = 0$ *if and only if* $f(t) = c, c \in \mathbb{R}$, *for all* $t \in [a,b]_{q,\omega}$.

Proof The implication "$\Leftarrow$" is obvious. We prove "$\Rightarrow$". We begin noting that

$$\underbrace{\int_a^b f(t)\,D_{q,\omega}\left[\eta\right](t)\,d_{q,\omega}t}_{=0} = \underbrace{f(t)\,\eta(t)\Big|_a^b}_{=0} - \int_a^b D_{q,\omega}\left[f\right](t)\,\eta^{\sigma}(t)\,d_{q,\omega}t.$$

Hence,

$$\int_a^b D_{q,\omega}\left[f\right](t)\,\eta(qt + \omega)\,d_{q,\omega}t = 0$$

for any $\eta \in \mathcal{Y}^1([a, b], \mathbb{R})$ such that $\eta(a) = \eta(b) = 0$. We need to prove that, for some $c \in \mathbb{R}$, $f(t) = c$ for all $t \in [a, b]_{q,\omega}$, that is, $D_{q,\omega}[f](t) = 0$ for all $t \in [a, b]_{q,\omega}$. Suppose, by contradiction, that there exists $p \in [a, b]_{q,\omega}$ such that $D_{q,\omega}[f](p) \neq 0$.

(1) If $p \neq \omega_0$, then $p = q^k a + \omega[k]_q$ or $p = q^k b + \omega[k]_q$ for some $k \in \mathbb{N}_0$. Observe that $a(1-q) - \omega$ and $b(1-q) - \omega$ cannot vanish simultaneously.

(a) Suppose that $a(1-q) - \omega \neq 0$ and $b(1-q) - \omega \neq 0$. In this case we can assume, without loss of generality, that $p = q^k a + \omega[k]_q$ and we can define

$$\eta(t) = \begin{cases} D_{q,\omega}[f](q^k a + \omega[k]_q) & \text{if } t = q^{k+1}a + \omega[k+1]_q \\ 0 & \text{otherwise.} \end{cases}$$

Then,

$$\int_a^b D_{q,\omega}[f](t) \cdot \eta(qt + \omega) \, d_{q,\omega}t$$

$$= -(a(1-q) - \omega) q^k D_{q,\omega}[f]\left(q^k a + \omega[k]_q\right) \cdot D_{q,\omega}[f]\left(q^k a + \omega[k]_q\right) \neq 0,$$

which is a contradiction.

(b) If $a(1-q) - \omega \neq 0$ and $b(1-q) - \omega = 0$, then $b = \omega_0$. Since $q^k \omega_0 + \omega[k]_q = \omega_0$ for all $k \in \mathbb{N}_0$, then $p \neq q^k b + \omega[k]_q \ \forall k \in \mathbb{N}_0$ and, therefore,

$$p = q^k a + \omega[k]_{q,\omega} \text{ for some } k \in \mathbb{N}_0.$$

Repeating the proof of (a) we obtain again a contradiction.

(c) If $a(1-q) - \omega = 0$ and $b(1-q) - \omega \neq 0$, then the proof is similar to (b).

(2) If $p = \omega_0$ then, without loss of generality, we can assume $D_{q,\omega}[f](\omega_0) > 0$. Since

$$\lim_{n \to +\infty} \left(q^n a + \omega[k]_q\right) = \lim_{n \to +\infty} \left(q^n b + \omega[k]_q\right) = \omega_0$$

(see Aldwoah (2009)) and $D_{q,\omega}[f]$ is continuous at ω_0, then

$$\lim_{n \to +\infty} D_{q,\omega}[f]\left(q^n a + \omega[k]_q\right) = \lim_{n \to +\infty} D_{q,\omega}[f]\left(q^n b + \omega[k]_q\right)$$

$$= D_{q,\omega}[f](\omega_0) > 0.$$

Thus, there exists $N \in \mathbb{N}$ such that for all $n \geq N$ one has

$$D_{q,\omega}[f]\left(q^n a + \omega[k]_q\right) > 0 \text{ and } D_{q,\omega}[f]\left(q^n b + \omega[k]_q\right) > 0.$$

(a) If $\omega_0 \neq a$ and $\omega_0 \neq b$, then we can define

$$\eta\left(t\right) = \begin{cases} D_{q,\omega}\left[f\right]\left(q^{N}b + \omega\left[N\right]_{q}\right) \ if \ t = q^{N+1}a + \omega\left[N+1\right]_{q} \\ D_{q,\omega}\left[f\right]\left(q^{N}a + \omega\left[N\right]_{q}\right) \ if \ t = q^{N+1}b + \omega\left[N+1\right]_{q} \\ 0 \hspace{4cm} \text{otherwise.} \end{cases}$$

Hence,

$$\int_{a}^{b} D_{q,\omega}\left[f\right]\left(t\right)\eta\left(qt + \omega\right) d_{q,\omega}t$$
$$= \left(b - a\right)\left(1 - q\right)q^{N}D_{q,\omega}\left[f\right]\left(q^{N}b + \omega\left[N\right]_{q}\right) \cdot D_{q\omega}\left[f\right]\left(q^{N}a + \omega\left[N\right]_{q}\right) \neq 0,$$

which is a contradiction.

(b) If $\omega_{0} = b$, then we define

$$\eta\left(t\right) = \begin{cases} D_{q,\omega}\left[f\right]\left(\omega_{0}\right) \ if \ t = q^{N+1}a + \omega\left[N+1\right]_{q} \\ 0 \hspace{4cm} \text{otherwise.} \end{cases}$$

Therefore,

$$\int_{a}^{b} D_{q,\omega}\left[f\right]\left(t\right)\eta\left(qt + \omega\right) d_{q,\omega}t$$
$$= -\int_{\omega_{0}}^{a} D_{q,\omega}\left[f\right]\left(t\right)\eta\left(qt + \omega\right) d_{q,\omega}t$$
$$= -\left(a\left(1 - q\right) - \omega\right)q^{N}D_{q,\omega}\left[f\right]\left(q^{N}a + \omega\left[k\right]_{q}\right) \cdot D_{q,\omega}\left[f\right]\left(\omega_{0}\right) \neq 0,$$

which is a contradiction.

(c) When $\omega_{0} = a$, the proof is similar to (b).

Lemma 2.42 (Fundamental lemma of Hahn's variational calculus) *Let $f, g \in \mathcal{Y}^{1}\left(\left[a, b\right], \mathbb{R}\right).$*

If

$$\int_{a}^{b}\left(f\left(t\right)\eta^{\sigma}\left(t\right) + g\left(t\right)D_{q,\omega}\left[\eta\right]\left(t\right)\right) d_{q,\omega}t = 0$$

for all $\eta \in \mathcal{Y}^{1}\left(\left[a, b\right], \mathbb{R}\right)$ such that $\eta\left(a\right) = \eta\left(b\right) = 0$, then

$$D_{q,\omega}\left[g\right]\left(t\right) = f\left(t\right) \ \forall t \in \left[a, b\right]_{q,\omega}.$$

Proof Define the function A by

$$A(t) := \int_{\omega_0}^{t} f(\tau)\, d_{q,\omega}\tau.$$

Then $D_{q,\omega}[A](t) = f(t)$ for all $t \in [a,b]$ and

$$\int_{a}^{b} A(t)\, D_{q,\omega}[\eta](t)\, d_{q,\omega}t = A(t)\eta(t)\Big|_{a}^{b} - \int_{a}^{b} D_{q,\omega}[A](t)\, \eta^{\sigma}(t)\, d_{q,\omega}t$$

$$= -\int_{a}^{b} D_{q,\omega}[A](t)\, \eta^{\sigma}(t)\, d_{q,\omega}t$$

$$= -\int_{a}^{b} f(t)\, \eta^{\sigma}(t)\, d_{q,\omega}t.$$

Hence,

$$\int_{a}^{b} \left(f(t)\eta^{\sigma}(t) + g(t)\, D_{q,\omega}[\eta](t) \right) d_{q,\omega}t = 0$$

$$\Leftrightarrow \int_{a}^{b} (-A(t) + g(t))\, D_{q,\omega}[\eta](t)\, d_{q,\omega}t = 0.$$

By Lemma 2.41 there is a $c \in \mathbb{R}$ such that $-A(t) + g(t) = c$ for all $t \in [a,b]_{q,\omega}$. Hence $D_{q,\omega}[A](t) = D_{q,\omega}[g](t)$ for $t \in [a,b]_{q,\omega}$, which provides the desired result: $D_{q,\omega}[g](t) = f(t)\ \forall t \in [a,b]_{q,\omega}$.

We are now in conditions to deduce the higher-order fundamental Lemma of Hahn's quantum variational calculus.

Lemma 2.43 (Higher-order fundamental lemma of Hahn's variational calculus) *Let $f_0, f_1, \ldots, f_r \in \mathcal{Y}^1([a,b], \mathbb{R})$. If*

$$\int_{a}^{b} \left(\sum_{i=0}^{r} f_i(t)\, D_{q,\omega}^{i}\left[\eta^{\sigma^{r-i}}\right](t) \right) d_{q,\omega}t = 0$$

for any variation η, then

$$\sum_{i=0}^{r} (-1)^{i} \left(\frac{1}{q}\right)^{\frac{(i-1)i}{2}} D_{q,\omega}^{i}[f_i](t) = 0$$

for all $t \in [a,b]_{q,\omega}$.

Proof We proceed by mathematical induction. If $r = 1$ the result is true by Lemma 2.42. Assume that

$$\int_a^b \left(\sum_{i=0}^{r+1} f_i(t) D_{q,\omega}^i \left[\eta^{\sigma^{r+1-i}} \right](t) \right) d_{q,\omega}t = 0$$

for all functions η such that $\eta(a) = \eta(b) = 0, ..., D_{q,\omega}^r[\eta](a) = D_{q,\omega}^r[\eta](b) = 0$.
Note that

$$\int_a^b f_{r+1}(t) D_{q,\omega}^{r+1}[\eta](t) d_{q,\omega}t$$

$$= f_{r+1}(t) D_{q,\omega}^r[\eta](t) \Big|_a^b - \int_a^b D_{q,\omega}[f_{r+1}](t) \left(D_{q,\omega}^r[\eta] \right)^\sigma (t) d_{q,\omega}t$$

$$= - \int_a^b D_{q,\omega}[f_{r+1}](t) \left(D_{q,\omega}^r[\eta] \right)^\sigma (t) d_{q,\omega}t$$

and, by Lemma 2.39,

$$\int_a^b f_{r+1}(t) D_{q,\omega}^{r+1}[\eta](t) d_{q,\omega}t = - \int_a^b D_{q,\omega}[f_{r+1}](t) \left(\frac{1}{q} \right)^r D_{q,\omega}^r[\eta^\sigma](t) d_{q,\omega}t.$$

Therefore,

$$\int_a^b \left(\sum_{i=0}^{r+1} f_i(t) D_{q,\omega}^i \left[\eta^{\sigma^{r+1-i}} \right](t) \right) d_{q,\omega}t$$

$$= \int_a^b \left(\sum_{i=0}^r f_i(t) D_{q,\omega}^i \left[\eta^{\sigma^{r+1-i}} \right](t) \right) d_{q,\omega}t$$

$$\qquad - \int_a^b D_{q,\omega}[f_{r+1}](t) \left(\frac{1}{q} \right)^r D_{q,\omega}^r[\eta^\sigma](t) d_{q,\omega}t$$

$$= \int_a^b \left[\sum_{i=0}^{r-1} f_i(t) D_{q,\omega}^i \left[(\eta^\sigma)^{\sigma^{r-i}} \right](t) d_{q,\omega}t \right.$$

$$\left. + \left(f_r - \left(\frac{1}{q} \right)^r D_{q,\omega}[f_{r+1}] \right)(t) D_{q,\omega}^r[\eta^\sigma](t) \right] d_{q,\omega}t.$$

By Lemma 2.40, η^σ is a variation. Hence, using the induction hypothesis,

$$\sum_{i=0}^{r-1} (-1)^i \left(\frac{1}{q} \right)^{\frac{(i-1)i}{2}} D_{q,\omega}^i[f_i](t)$$

$$+ (-1)^r \left(\frac{1}{q} \right)^{\frac{(r-1)r}{2}} D_{q,\omega}^r \left[\left(f_r - \frac{1}{q^r} D_{q,\omega}[f_{r+1}] \right) \right](t)$$

$$= \sum_{i=0}^{r-1} (-1)^i \left(\frac{1}{q}\right)^{\frac{(i-1)i}{2}} D_{q,\omega}^i [f_i](t) + (-1)^r \left(\frac{1}{q}\right)^{\frac{(r-1)r}{2}} D_{q,\omega}^r [f_r](t)$$

$$+ (-1)^{r+1} \left(\frac{1}{q}\right)^{\frac{(r-1)r}{2}} \frac{1}{q^r} D_{q,\omega}^r \left[D_{q,\omega}[f_{r+1}]\right](t)$$

$$= 0$$

for all $t \in [a,b]_{q,\omega}$, which leads to

$$\sum_{i=0}^{r+1} (-1)^i \left(\frac{1}{q}\right)^{\frac{(i-1)i}{2}} D_{q,\omega}^i [f_i](t) = 0, \ t \in [a,b]_{q,\omega}.$$

2.7.2 Higher-order Hahn's Quantum Euler–Lagrange Equation

For a variation η and an admissible function y, we define the function $\phi : (-\bar{\epsilon}, \bar{\epsilon}) \to \mathbb{R}$ by $\phi(\epsilon) = \phi(\epsilon, y, \eta) := \mathcal{L}[y + \epsilon\eta]$. The first variation of the variational problem (2.7) is defined by $\delta\mathcal{L}[y, \eta] := \phi'(0)$. Observe that

$$\mathcal{L}[y + \epsilon\eta] = \int_a^b L\Big(t, y^{\sigma^r}(t) + \epsilon\eta^{\sigma^r}(t), D_{q,\omega}\Big[y^{\sigma^{r-1}}\Big](t) + \epsilon D_{q,\omega}\Big[\eta^{\sigma^{r-1}}\Big](t),$$

$$\ldots, D_{q,\omega}^r [y](t) + \epsilon D_{q,\omega}^r [\eta](t)\Big) d_{q,\omega}t$$

$$= \mathcal{L}_b[y + \epsilon\eta] - \mathcal{L}_a[y + \epsilon\eta]$$

with

$$\mathcal{L}_\xi[y + \epsilon\eta] = \int_{\omega_0}^\xi L\Big(t, y^{\sigma^r}(t) + \epsilon\eta^{\sigma^r}(t), D_{q,\omega}\Big[y^{\sigma^{r-1}}\Big](t) + \epsilon D_{q,\omega}\Big[\eta^{\sigma^{r-1}}\Big](t),$$

$$\ldots, D_{q,\omega}^r [y](t) + \epsilon D_{q,\omega}^r [\eta](t)\Big) d_{q,\omega}t,$$

$\xi \in \{a, b\}$. Therefore,

$$\delta\mathcal{L}[y, \eta] = \delta\mathcal{L}_b[y, \eta] - \delta\mathcal{L}_a[y, \eta]. \tag{2.29}$$

Considering (2.29), the following is a direct consequence of Lemma 2.15:

Lemma 2.44 *For a variation η and an admissible function y, let*

$$g(t, \epsilon) := L\bigg(t, y^{\sigma^r}(t) + \epsilon\eta^{\sigma^r}(t), D_{q,\omega}\Big[y^{\sigma^{r-1}}\Big](t) + \epsilon D_{q,\omega}\Big[\eta^{\sigma^{r-1}}\Big](t),$$

$$\dots, D_{q,\omega}^r[y](t) + \epsilon D_{q,\omega}^r[\eta](t)\bigg),$$

$\epsilon \in (-\bar{\epsilon}, \bar{\epsilon})$. *Assume that:*
(1) $g(t, \cdot)$ is differentiable at 0 uniformly in $t \in [a, b]_{q,\omega}$;
(2) $\mathcal{L}_a[y + \epsilon\eta] = \int_{\omega_0}^a g(t, \epsilon)\,d_{q,\omega}t$ and $\mathcal{L}_b[y + \epsilon\eta] = \int_{\omega_0}^b g(t, \epsilon)\,d_{q,\omega}t$ exist for
$\epsilon \approx 0$;
(3) $\int_{\omega_0}^a \partial_2 g(t, 0)\,d_{q,\omega}t$ and $\int_{\omega_0}^b \partial_2 g(t, 0)\,d_{q,\omega}t$ exist.
Then

$$\phi'(0) = \delta\mathcal{L}[y, \eta]$$

$$= \int_a^b \bigg(\sum_{i=0}^r \partial_{i+2}L\Big(t, y^{\sigma^r}(t), D_{q,\omega}\Big[y^{\sigma^{r-1}}\Big](t), \dots, D_{q,\omega}^r[y](t)\Big)$$

$$\cdot D_{q,\omega}^i\Big[\eta^{\sigma^{r-i}}\Big](t)\bigg)d_{q,\omega}t,$$

where $\partial_i L$ denotes the partial derivative of L with respect to its ith argument.

The following result gives a necessary condition of Euler–Lagrange type for an admissible function to be a local extremizer for (2.7).

Theorem 2.45 (The higher-order Hahn quantum Euler–Lagrange equation)
Under hypotheses (H1)–(H3) and conditions (1)–(3) of Lemma 2.44 on the Lagrangian L, if $y_* \in \mathcal{Y}^r$ is a local extremizer for problem (2.7), then y_* satisfies the q, ω-Euler–Lagrange equation

$$\sum_{i=0}^r (-1)^i \left(\frac{1}{q}\right)^{\frac{(i-1)i}{2}} D_{q,\omega}^i[\partial_{i+2}L]\Big(t, y^{\sigma^r}(t), D_{q,\omega}\Big[y^{\sigma^{r-1}}\Big](t),$$

$$\dots, D_{q,\omega}^r[y](t)\Big) = 0 \qquad\qquad (2.30)$$

for all $t \in [a, b]_{q,\omega}$.

Proof Let y_* be a local extremizer for problem (2.7) and η a variation. Define $\phi : (-\bar{\epsilon}, \bar{\epsilon}) \to \mathbb{R}$ by $\phi(\epsilon) := \mathcal{L}[y_* + \epsilon\eta]$. A necessary condition for y_* to be an extremizer is given by $\phi'(0) = 0$. By Lemma 2.44 we conclude that

$$\int_a^b \left(\sum_{i=0}^r \partial_{i+2} L \left(t, y^{\sigma^r}(t), D_{q,\omega} \left[y^{\sigma^{r-1}} \right](t), \dots, D_{q,\omega}^r [y](t) \right) \right.$$

$$\left. \cdot D_{q,\omega}^i \left[\eta^{\sigma^{r-i}} \right](t) \right) d_{q,\omega} t = 0$$

and (2.30) follows from Lemma 2.43.

Remark 2.46 In practical terms the hypotheses of Theorem 2.45 are not so easy to verify *a priori*. One can, however, assume that all hypotheses are satisfied and apply the q, ω-Euler–Lagrange equation (2.30) heuristically to obtain a *candidate*. If such a candidate is, or not, a solution to problem (2.7) is a different question that always requires further analysis (see an example in Sect. 2.7.3).

When $\omega = 0$ one obtains from (2.30) the higher-order q-Euler–Lagrange equation:

$$\sum_{i=0}^r (-1)^i \left(\frac{1}{q} \right)^{\frac{(i-1)i}{2}} D_q^i \left[\partial_{i+2} L \right] \left(t, y^{\sigma^r}(t), D_q \left[y^{\sigma^{r-1}} \right](t), \dots, D_q^r [y](t) \right) = 0$$

for all $t \in \{aq^n : n \in \mathbb{N}_0\} \cup \{bq^n : n \in \mathbb{N}_0\} \cup \{0\}$. The higher-order h-Euler–Lagrange equation is obtained from (2.30) taking the limit $q \to 1$:

$$\sum_{i=0}^r (-1)^i \Delta_h^i \left[\partial_{i+2} L \right] \left(t, y^{\sigma^r}(t), \Delta_h \left[y^{\sigma^{r-1}} \right](t), \dots, \Delta_h^r [y](t) \right) = 0$$

for all $t \in \{a + nh : n \in \mathbb{N}_0\} \cup \{b + nh : n \in \mathbb{N}_0\}$. The classical Euler–Lagrange equation (van Brunt 2004) is recovered when $(\omega, q) \to (0, 1)$:

$$\sum_{i=0}^r (-1)^i \frac{d^i}{dt^i} \partial_{i+2} L \left(t, y(t), y'(t), \dots, y^{(r)}(t) \right) = 0$$

for all $t \in [a, b]$.

We now illustrate the usefulness of our Theorem 2.45 by means of an example that is not covered by previous available results in the literature.

2.7.3 An Example

Let $q = \frac{1}{2}$ and $\omega = \frac{1}{2}$. Consider the following problem:

$$\mathcal{L}[y] = \int_{-1}^1 \left(y^\sigma(t) + \frac{1}{2} \right)^2 \left((D_{q,\omega} [y](t))^2 - 1 \right)^2 d_{q,\omega} t \longrightarrow \min \qquad (2.31)$$

over all $y \in \mathcal{Y}^1$ satisfying the boundary conditions

$$y(-1) = 0 \quad \text{and} \quad y(1) = -1. \tag{2.32}$$

This is an example of problem (2.7) with $r = 1$. Our q, ω-Euler–Lagrange equation (2.30) takes the form

$$D_{q,\omega}[\partial_3 L]\left(t, y^\sigma(t), D_{q,\omega}[y](t)\right) = \partial_2 L\left(t, y^\sigma(t), D_{q,\omega}[y](t)\right).$$

Therefore, we look for an admissible function y_* of (2.31)–(2.32) satisfying

$$D_{q,\omega}\left[4\left(y^\sigma + \frac{1}{2}\right)^2\left((D_{q,\omega}[y])^2 - 1\right)D_{q,\omega}[y]\right](t)$$

$$= 2\left(y^\sigma(t) + \frac{1}{2}\right)\left((D_{q,\omega}[y](t))^2 - 1\right) \quad (2.33)$$

for all $t \in [-1, 1]_{q,\omega}$. It is easy to see that

$$y_*(t) = \begin{cases} -t & \text{if } t \in (-1, 0) \cup (0, 1] \\ 0 & \text{if } t = -1 \\ 1 & \text{if } t = 0 \end{cases}$$

is an admissible function for (2.31)–(2.32) with

$$D_{q,\omega}[y_*](t) = \begin{cases} -1 & \text{if } t \in (-1, 0) \cup (0, 1] \\ 1 & \text{if } t = -1 \\ -3 & \text{if } t = 0, \end{cases}$$

satisfying the q, ω-Euler–Lagrange equation (2.33). We now prove that the *candidate* y_* is indeed a minimizer for (2.31)–(2.32). Note that here $\omega_0 = 1$ and, by Lemma 2.11 and item (3) of Theorem 2.10,

$$\mathcal{L}[y] = \int_{-1}^{1}\left(y^\sigma(t) + \frac{1}{2}\right)^2\left((D_{q,\omega}[y](t))^2 - 1\right)^2 d_{q,\omega}t \geq 0 \tag{2.34}$$

for all admissible functions $y \in \mathcal{Y}^1([-1, 1], \mathbb{R})$. Since $\mathcal{L}[y_*] = 0$, we conclude that y_* is a minimizer for problem (2.31)–(2.32).

It is worth mentioning that the minimizer y_* of (2.31)–(2.32) is not continuous while the classical calculus of variations (van Brunt 2004), the calculus of variations on time scales (Ferreira and Torres 2008; Malinowska and Torres 2009; Martins and Torres 2009), or the nondifferentiable scale variational calculus (Almeida and Torres 2009a, 2010a; Cresson et al. 2009), deal with functions which are necessarily continuous. As an open question, we pose the problem of determining conditions on the data of problem (2.7) assuring, *a priori*, the minimizer to be regular.

2.8 Generalized Transversality Conditions

The main purpose of this section is to generalize the Hahn calculus of variations (Malinowska and Torres 2010c) by considering the following q, ω-variational problem:

$$\mathcal{L}[y] = \int_a^b L\left(t, y\left(qt + \omega\right), D_{q,\omega}\left[y\right](t), y(a), y(b)\right) d_{q,\omega} t \longrightarrow \text{extr.} \qquad (2.35)$$

In Sect. 2.8.1 we obtain the Euler–Lagrange equation for problem (2.35) in the class of functions $y \in \mathcal{Y}^1$ satisfying the boundary conditions

$$y(a) = \alpha \quad \text{and} \quad y(b) = \beta \qquad (2.36)$$

for some fixed $\alpha, \beta \in \mathbb{R}$. The transversality conditions for problem (2.35) are obtained in Sect. 2.8.2. In Sect. 2.8.3 we prove necessary optimality conditions for isoperimetric problems. A sufficient optimality condition under an appropriate convexity assumption is given in Sect. 2.8.4.

In the sequel we assume that the Lagrangian L satisfies the following hypotheses:

(H1) $(u_0, \ldots, u_3) \to L(t, u_0, \ldots, u_3)$ is a $C^1(\mathbb{R}^4, \mathbb{R})$ function for any $t \in [a, b]$;
(H2) $t \to L(t, y(qt + \omega), D_{q,\omega}[y](t), y(a), y(b))$ is continuous at ω_0 for any $y \in \mathcal{Y}^1$;
(H3) functions $t \to \partial_{i+2} L(t, y(qt + \omega), D_{q,\omega}[y](t), y(a), y(b)), i = 0, \ldots, 3$ belong to $\mathcal{Y}^1$ for all $y \in \mathcal{Y}^1$.

In order to simplify expressions, we introduce the operator $\{\cdot\}$ defined in the following way:

$$\{y\}(t, a, b) := (t, y(qt + \omega), D_{q,\omega}[y](t), y(a), y(b))$$

where $y \in \mathcal{Y}^1$.

The following lemma can be obtained similar to Lemma 2.15.

Lemma 2.47 *For fixed $y, h \in \mathcal{Y}^1$ let*

$$g(t, \varepsilon) = L(t, y(qt + \omega) + \varepsilon h(qt + \omega), D_{q,\omega}[y](t)$$
$$+ \varepsilon D_{q,\omega}[h](t), y(a) + \varepsilon h(a), y(b) + \varepsilon h(b))$$

for $\varepsilon \in] -\bar{\varepsilon}, \bar{\varepsilon}[$, for some $\bar{\varepsilon} > 0$, i.e., $g(t, \varepsilon) = L\{y + \varepsilon h\}(t, a, b)$. Assume that:

(i) *$g(t, \cdot)$ is differentiable at 0 uniformly in $t \in [a, b]_{q,\omega}$;*

(ii) $\mathcal{L}_a[y + \varepsilon h] = \displaystyle\int_{\omega_0}^{a} g(t, \epsilon) \, d_{q,\omega} t$ and $\mathcal{L}_b[y + \varepsilon h] = \displaystyle\int_{\omega_0}^{b} g(t, \epsilon) \, d_{q,\omega} t$ exist for $\varepsilon \approx 0$;

(iii) $\displaystyle\int_{\omega_0}^{a} \partial_2 g(t, 0) d_{q,\omega} t$ and $\displaystyle\int_{\omega_0}^{b} \partial_2 g(t, 0) d_{q,\omega} t$ exist.

Then,

$$\delta\mathcal{L}[y, h] = \int_a^b \Big(\partial_2 L\{y\}(t, a, b) \cdot h(qt + \omega) + \partial_3 L\{y\}(t, a, b) \cdot D_{q,\omega}[h](t)$$
$$+ \partial_4 L\{y\}(t, a, b) \cdot h(a) + \partial_5 L\{y\}(t, a, b) \cdot h(b) \Big) d_{q,\omega} t.$$

2.8.1 The Hahn Quantum Euler–Lagrange Equation

In the following theorem, we give the Euler–Lagrange equation for problem (2.35)–(2.36).

Theorem 2.48 (Necessary optimality condition to (2.35)–(2.36)) *Under hypotheses (H1)–(H3) and conditions (i)–(iii) of Lemma 2.47 on the Lagrangian L, if $\tilde{y}$ is a local minimizer or local maximizer to problem (2.35)–(2.36), then $\tilde{y}$ satisfies the Euler–Lagrange equation*

$$\partial_2 L\{y\}(t, a, b) - D_{q,\omega}[\partial_3 L]\{y\}(t, a, b) = 0 \qquad (2.37)$$

for all $t \in [a, b]_{q,\omega}$.

Proof Suppose that $\mathcal{L}$ has a local extremum at $\tilde{y}$. Let h be any admissible variation and define a function $\phi :] - \bar{\varepsilon}, \bar{\varepsilon}[\to \mathbb{R}$ by $\phi(\varepsilon) = \mathcal{L}[\tilde{y} + \varepsilon h]$. A necessary condition for $\tilde{y}$ to be an extremizer is given by $\phi'(0) = 0$. Note that

$$\phi'(0) = \int_a^b \Big(\partial_2 L\{\tilde{y}\}(t, a, b) \cdot h(qt + \omega) + \partial_3 L\{\tilde{y}\}(t, a, b) \cdot D_{q,\omega}[h](t)$$
$$+ \partial_4 L\{\tilde{y}\}(t, a, b) \cdot h(a) + \partial_5 L\{\tilde{y}\}(t, a, b) \cdot h(b) \Big) d_{q,\omega} t.$$

Since $h(a) = h(b) = 0$, then

$$\phi'(0) = \int_a^b \Big(\partial_2 L\{\tilde{y}\}(t, a, b) \cdot h(qt + \omega) + \partial_3 L\{\tilde{y}\}(t, a, b) \cdot D_{q,\omega}[h](t) \Big) d_{q,\omega} t.$$

Integration by parts gives

$$\int_a^b \partial_3 L\{\tilde{y}\}(t, a, b) \cdot D_{q,\omega}[h](t)d_{q,\omega}t = \Big[\partial_3 L\{\tilde{y}\}(t, a, b) \cdot h(t)\Big]_a^b$$

$$- \int_a^b D_{q,\omega}[\partial_3 L]\{\tilde{y}\}(t, a, b) \cdot h(qt + \omega)d_{q,\omega}t$$

and since $h(a) = h(b) = 0$, then

$$\phi'(0) = 0 \Leftrightarrow \int_a^b \Big(\partial_2 L\{\tilde{y}\}(t, a, b) - D_{q,\omega}[\partial_3 L]\{\tilde{y}\}(t, a, b)\Big) \cdot h(qt + \omega)d_{q,\omega}t = 0.$$

Thus, by Lemma 2.16, we have

$$\partial_2 L\{\tilde{y}\}(t, a, b) - D_{q,\omega}[\partial_3 L]\{\tilde{y}\}(t, a, b) = 0$$

for all $t \in [a, b]_{q,\omega}$.

Remark 2.49 Under appropriate conditions, when $(\omega, q) \to (0, 1)$, we obtain a corresponding result in the classical context of the calculus of variations (Cruz et al. 2010) (see also Malinowska and Torres (2010b)):

$$\frac{d}{dt}\partial_3 L(t, y(t), y'(t), y(a), y(b)) = \partial_2 L(t, y(t), y'(t), y(a), y(b)).$$

Remark 2.50 In the basic problem of the calculus of variations, L does not depend on $y(a)$ and $y(b)$, and equation (2.37) reduces to the Hahn quantum Euler–Lagrange equation (2.4).

2.8.2 Natural Boundary Conditions

The following theorem provides necessary optimality conditions for problem (2.35).

Theorem 2.51 (Natural boundary conditions to (2.35)) *Under hypotheses (H1)– (H3) and conditions (i)–(iii) of Lemma 2.47 on the Lagrangian L, if $\tilde{y}$ is a local minimizer or local maximizer to problem (2.35), then $\tilde{y}$ satisfies the Euler–Lagrange equation (2.37) and*

1. *if $y(a)$ is free, then the natural boundary condition*

$$\partial_3 L\{\tilde{y}\}(a, a, b) = \int_a^b \partial_4 L\{\tilde{y}\}(t, a, b)d_{q,\omega}t \qquad (2.38)$$

   holds;
2. *if $y(b)$ is free, then the natural boundary condition*

$$\partial_3 L\{\tilde{y}\}(b, a, b) = -\int_a^b \partial_5 L\{\tilde{y}\}(t, a, b)d_{q,\omega}t \tag{2.39}$$

holds.

Proof Suppose that $\tilde{y}$ is a local minimizer (resp. maximizer) to problem (2.35). Let h be any $\mathcal{Y}^1$ function. Define a function $\phi :\,]-\bar{\varepsilon}, \bar{\varepsilon}[\to \mathbb{R}$ by $\phi(\varepsilon) = \mathcal{L}[\tilde{y} + \varepsilon h]$. It is clear that a necessary condition for $\tilde{y}$ to be an extremizer is given by $\phi'(0) = 0$. From the arbitrariness of h and using similar arguments as the ones used in the proof of Theorem 2.48, it can be proved that $\tilde{y}$ satisfies the Euler–Lagrange equation (2.37).

1. Suppose now that $y(a)$ is free. If $y(b) = \beta$ is given, then $h(b) = 0$; if $y(b)$ is free, then we restrict ourselves to those h for which $h(b) = 0$. Therefore,

$$\begin{aligned}
0 &= \phi'(0) \\
&= \int_a^b \left(\partial_2 L\{\tilde{y}\}(t, a, b) - D_{q,\omega}[\partial_3 L]\{\tilde{y}\}(t, a, b) \right) \cdot h(qt + w)d_{q,\omega}t \quad (2.40) \\
&\quad + \left(\int_a^b \partial_4 L\{\tilde{y}\}(t, a, b)d_{q,\omega}t - \partial_3 L\{\tilde{y}\}(a, a, b) \right) \cdot h(a) = 0.
\end{aligned}$$

Using the Euler–Lagrange equation (2.37) into (2.40) we obtain

$$\left(\int_a^b \partial_4 L\{\tilde{y}\}(t, a, b)d_{q,\omega}t - \partial_3 L\{\tilde{y}\}(a, a, b) \right) \cdot h(a) = 0.$$

From the arbitrariness of h it follows that

$$\partial_3 L\{\tilde{y}\}(a, a, b) = \int_a^b \partial_4 L\{\tilde{y}\}(t, a, b)d_{q,\omega}t.$$

2. Suppose now that $y(b)$ is free. If $y(a) = \alpha$, then $h(a) = 0$; if $y(a)$ is free, then we restrict ourselves to those h for which $h(a) = 0$. Thus,

$$\begin{aligned}
0 &= \phi'(0) \\
&= \int_a^b \left(\partial_2 L\{\tilde{y}\}(t, a, b) - D_{q,\omega}[\partial_3 L]\{\tilde{y}\}(t, a, b) \right) \cdot h(qt + w)d_{q,\omega}t \quad (2.41) \\
&\quad + \left(\int_a^b \partial_5 L\{\tilde{y}\}(t, a, b)d_{q,\omega}t + \partial_3 L\{\tilde{y}\}(b, a, b) \right) \cdot h(b) = 0.
\end{aligned}$$

Using the Euler–Lagrange equation (2.37) into (2.41), and from the arbitrariness of h, it follows that

$$\partial_3 L\{\tilde{y}\}(b, a, b) = -\int_a^b \partial_5 L\{\tilde{y}\}(t, a, b)d_{q,\omega}t.$$

In the case where L does not depend on $y(a)$ and $y(b)$, under appropriate assumptions on the Lagrangian L, we obtain the following result.

Corollary 2.52 *If $\tilde{y}$ is a local minimizer or local maximizer to problem*

$$\mathcal{L}[y] = \int_a^b L\{\tilde{y}\}(t)d_{q,\omega}t \longrightarrow extr$$

then $\tilde{y}$ satisfies the Euler–Lagrange equation

$$\partial_2 L\{\tilde{y}\}(t) - D_{q,\omega}[\partial_3 L]\{\tilde{y}\}(t) = 0$$

for all $t \in [a, b]_{q,\omega}$, and

1. *if $y(a)$ is free, then the natural boundary condition*

$$\partial_3 L\{\tilde{y}\}(a) = 0 \tag{2.42}$$

   holds;
2. *if $y(b)$ is free, then the natural boundary condition*

$$\partial_3 L\{\tilde{y}\}(b) = 0 \tag{2.43}$$

   holds.

Remark 2.53 Under appropriate conditions, when $(\omega, q) \to (0, 1)$ equations (2.42) and (2.43) reduce to the well-known natural boundary conditions for the basic problem of the calculus of variations

$$\partial_3 L(a, \tilde{y}(a), \tilde{y}'(a)) = 0 \quad \text{and} \quad \partial_3 L(b, \tilde{y}(b), \tilde{y}'(b)) = 0,$$

respectively.

2.8.3 Isoperimetric Problem

We now study the general Hahn quantum isoperimetric problem with an integral constraint. Both normal and abnormal extremizers are considered. The isoperimetric problem consists of minimizing or maximizing the functional

$$\mathcal{L}[y] = \int_a^b L\left(t, y\left(qt + \omega\right), D_{q,\omega}[y]\left(t\right), y(a), y(b)\right) d_{q,\omega}t \tag{2.44}$$

in the class of functions $y \in \mathcal{Y}^1$ satisfying the integral constraint

$$ J[y] = \int_a^b F\left(t, y\left(qt + \omega\right), D_{q,\omega}[y](t), y(a), y(b)\right) d_{q,\omega}t = \gamma \qquad (2.45) $$

for some $\gamma \in \mathbb{R}$.

Theorem 2.54 (Necessary optimality condition for normal extremizers to (2.44)–(2.45)) *Suppose that L and F satisfy hypotheses (H1)–(H3) and conditions (i)–(iii) of Lemma 2.47, and suppose that $\widetilde{y} \in \mathcal{Y}^1$ gives a local minimum or a local maximum to the functional $\mathcal{L}$ subject to the integral constraint (2.45). If $\widetilde{y}$ is not an extremal to J, then there exists a real λ such that $\widetilde{y}$ satisfies the equation*

$$ \partial_2 H\{y\}(t, a, b) - D_{q,\omega}[\partial_3 H]\{y\}(t, a, b) = 0 \qquad (2.46) $$

for all $t \in [a, b]_{q,\omega}$, where $H = L - \lambda F$ and

1. *if $y(a)$ is free, then the natural boundary condition*

$$ \partial_3 H\{\widetilde{y}\}(a, a, b) = \int_a^b \partial_4 H\{\widetilde{y}\}(t, a, b) d_{q,\omega}t \qquad (2.47) $$

   holds;
2. *if $y(b)$ is free, then the natural boundary condition*

$$ \partial_3 H\{\widetilde{y}\}(b, a, b) = -\int_a^b \partial_5 H\{\widetilde{y}\}(t, a, b) d_{q,\omega}t \qquad (2.48) $$

   holds.

Proof The proof is left to the reader. Hint: recall proofs of Theorem 2.26 and Theorem 2.51.

Introducing an extra multiplier λ_0 we can also deal with abnormal extremizers to the isoperimetric problem (2.44)–(2.45).

Theorem 2.55 (Necessary optimality condition for normal and abnormal extremizers to (2.44)–(2.45)) *Suppose that L and F satisfy hypotheses (H1)–(H3) and conditions (i)–(iii) of Lemma 2.47, and suppose that $\widetilde{y} \in \mathcal{Y}^1$ gives a local minimum or a local maximum to the functional $\mathcal{L}$ subject to the integral constraint (2.45). Then there exist two constants λ_0 and λ, not both zero, such that $\widetilde{y}$ satisfies the equation*

$$ \partial_2 H\{y\}(t, a, b) - D_{q,\omega}[\partial_3 H]\{y\}(t, a, b) = 0 \qquad (2.49) $$

for all $t \in [a, b]_{q,\omega}$, where $H = \lambda_0 L - \lambda F$ and

1. *if $y(a)$ is free, then the natural boundary condition*

$$ \partial_3 H\{\widetilde{y}\}(a, a, b) = \int_a^b \partial_4 H\{\widetilde{y}\}(t, a, b) d_{q,\omega}t \qquad (2.50) $$

holds;

2. *if $y(b)$ is free, then the natural boundary condition*

$$\partial_3 H\{\tilde{y}\}(b, a, b) = -\int_a^b \partial_5 H\{\tilde{y}\}(t, a, b)d_{q,\omega}t \qquad (2.51)$$

holds.

In the case where L and F do not depend on $y(a)$ and $y(b)$, under appropriate assumptions on Lagrangians L and F, we obtain the following result.

Corollary 2.56 *If $\tilde{y}$ is a local minimizer or local maximizer to the problem*

$$\mathcal{L}[y] = \int_a^b L\{y\}(t)d_{q,\omega}t \longrightarrow extr$$

subject to the integral constraint

$$\mathcal{J}[y] = \int_a^b F\{y\}(t)d_{q,\omega}t = \gamma$$

for some $\gamma \in \mathbb{R}$, then there exist two constants λ_0 and λ, not both zero, such that $\tilde{y}$ satisfies the following equation

$$\partial_2 H\{y\}(t) - D_{q,\omega}[\partial_3 H]\{y\}(t) = 0$$

for all $t \in [a, b]_{q,\omega}$, where $H = \lambda_0 L - \lambda F$ and

1. *if $y(a)$ is free, then the natural boundary condition*

$$\partial_3 H\{\tilde{y}\}(a) = 0$$

   holds;
2. *if $y(b)$ is free, then the natural boundary condition*

$$\partial_3 H\{\tilde{y}\}(b) = 0$$

   holds.

2.8.4 Sufficient Condition for Optimality

The following theorem gives sufficient optimality conditions for problem (2.35).

Theorem 2.57 *Let* $L(t, u_1, \ldots, u_4)$ *be jointly convex (respectively concave) in* $(u_1, \ldots, u_4)$. *If* $\tilde{y}$ *satisfies conditions* (2.37), (2.38) *and* (2.39), *then* $\tilde{y}$ *is a global minimizer (respectively maximizer) to problem* (2.35).

Proof The proof can be adapted from the proof of Theorem 2.29.

2.8.5 Illustrative Examples

We provide some examples in order to illustrate our results.

Example 2.58 Let $q \in]0, 1[$ and $\omega \geq 0$ be fixed real numbers, and I be an interval of $\mathbb{R}$ such that $\omega_0, 0, 1 \in I$. Consider the problem

$$\mathcal{L}[y] = \int_0^1 \left(y(qt + \omega) + \frac{1}{2}(D_{q,\omega}[y](t))^2 \right) d_{q,\omega}t \longrightarrow \min \qquad (2.52)$$

over all $y \in \mathcal{Y}^1$ satisfying the boundary condition $y(1) = 1$. If $\tilde{y}$ is a local minimizer to problem (2.52), then by Corollary 2.52 it satisfies the following conditions:

$$D_{q,\omega} D_{q,\omega}[\tilde{y}](t) = 1, \qquad (2.53)$$

for all $t \in \{\omega[n]_q : n \in \mathbb{N}_0\} \cup \{q^n + \omega[n]_q : n \in \mathbb{N}_0\} \cup \{\omega_0\}$ and

$$D_{q,\omega}[\tilde{y}](0) = 0. \qquad (2.54)$$

It is easy to verify that $\tilde{y}(t) = \frac{1}{q+1}t^2 - (\frac{\omega}{q+1} - c)t + d$, where $c, d \in \mathbb{R}$, is a solution to equation (2.53). Using the natural boundary condition (2.54) we obtain that $c = 0$. In order to determine d we use the fixed boundary condition $y(1) = 1$, and obtain that $d = \frac{q+\omega}{q+1}$. Hence

$$\tilde{y}(t) = \frac{1}{q+1}t^2 - \frac{\omega}{q+1}t + \frac{q+\omega}{q+1}$$

is a candidate to be a minimizer to problem (2.52). Moreover, since L is jointly convex, by Theorem 2.57, $\tilde{y}$ is a global minimizer to problem (2.52).

Example 2.59 Let $q \in]0, 1[$ and $\omega \geq 0$ be fixed real numbers, and I be an interval of $\mathbb{R}$ such that $\omega_0, 0, 1 \in I$. Consider the problem of minimizing

$$\mathcal{L}[y] = \int_0^1 \left(y(qt + \omega) + \frac{1}{2}(D_{q,\omega}[y](t))^2 + \gamma \frac{1}{2}(y(1) - 1)^2 + \lambda \frac{1}{2}y^2(0) \right) d_{q,\omega}t, \qquad (2.55)$$

where $\gamma, \lambda \in \mathbb{R}^+$. If $\widetilde{y}$ is a local minimizer to (2.55), then by Theorem 2.51 it satisfies the following conditions:

$$D_{q,\omega} D_{q,\omega}[\widetilde{y}](t) = 1, \tag{2.56}$$

for all $t \in \{\omega[n]_q : n \in \mathbb{N}_0\} \cup \{q^n + \omega[n]_q : n \in \mathbb{N}_0\} \cup \{\omega_0\}$, and

$$D_{q,\omega}[\widetilde{y}](0) = \int_0^1 \lambda \widetilde{y}(0) d_{q,\omega} t, \tag{2.57}$$

$$D_{q,\omega}[\widetilde{y}](1) = - \int_0^1 \gamma(\widetilde{y}(1) - 1) d_{q,\omega} t. \tag{2.58}$$

As in Example 2.58, $\widetilde{y}(t) = \frac{1}{q+1}t^2 - (\frac{\omega}{q+1} - c)t + d$, where $c, d \in \mathbb{R}$, is a solution to equation (2.56). In order to determine c and d we use the natural boundary conditions (2.57) and (2.58). This gives

$$\begin{aligned}
\widetilde{y}(t) = &\frac{1}{q+1}t^2 - \frac{\omega(\lambda + \gamma) - \lambda(\gamma - 1)(q + 1) + \gamma\lambda}{(q + 1)(\gamma + \lambda\gamma + \lambda)}t \\
&+ \frac{(\gamma - 1)(q + 1) - \gamma(1 - \omega)}{(q + 1)(\gamma + \lambda\gamma + \lambda)}
\end{aligned} \tag{2.59}$$

as a candidate to be a minimizer to (2.55). Moreover, since L is jointly convex, by Theorem 2.57 it is a global minimizer. The minimizer (2.59) is represented in Fig. 2.1 for fixed $\gamma = \lambda = 2, q = 0.99$ and different values of ω.

Fig. 2.1 The minimizer (2.59) of Example 2.59 for fixed $\gamma = \lambda = 2, q = 0.99$ and different values of ω

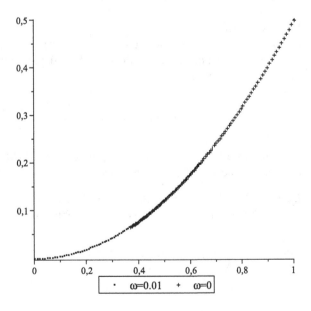

We note that in the limit, when $\gamma, \lambda \to +\infty$, $\widetilde{y}(t) = \frac{1}{q+1}t^2 + \frac{q}{q+1}t$ and coincides with the solution of the following problem with fixed initial and terminal points (see Example 2.32):

$$\mathcal{L}[y] = \int_0^1 \left(y(qt + \omega) + \frac{1}{2}(D_{q,\omega}[y](t))^2 \right) d_{q,\omega}t \longrightarrow \min$$

subject to the boundary conditions

$$y(0) = 0, \quad y(1) = 1.$$

Expression $\gamma\frac{1}{2}(y(1) - 1)^2 + \lambda\frac{1}{2}y^2(0)$ added to the Lagrangian $y(qt + \omega) + \frac{1}{2}(D_{q,\omega}[y](t))^2$ works like a penalty function when γ and λ go to infinity. The penalty function itself grows, and forces the merit function (2.55) to increase in value when the constraints $y(0) = 0$ and $y(1) = 1$ are violated, and causes no growth when constraints are fulfilled. The minimizer (2.59) is represented in Fig. 2.2 for fixed $q = 0.5, \omega = 1$ and different values of γ and λ.

Remark 2.60 Let

$$\mathcal{L}[y] = \int_0^1 \left(y(qt + \omega) + \frac{1}{2}(D_{q,\omega}[y](t))^2 \right) d_{q,\omega}t$$

and

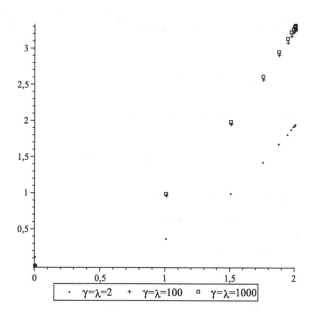

Fig. 2.2 The minimizer (2.59) of Example 2.59 for fixed $q = 0.5, \omega = 1$ and different values of γ and λ

$$\widetilde{y}_1(t) = \frac{1}{q+1}t^2 - \frac{\omega}{q+1}t + \frac{q+\omega}{q+1} \quad \text{and} \quad \widetilde{y}_2(t) = \frac{1}{q+1}t^2 + \frac{q}{q+1}t.$$

Comparing Example 2.58 and Example 2.59, we can conclude that

$$\mathcal{L}[\widetilde{y}_1] < \mathcal{L}[\widetilde{y}_2].$$

2.9 An Application Towards Economics

As the variables, that are usually considered and observed by the economist, are the outcome of a great number of decisions, taken by different operators at different points of time, it seems natural to look for new kinds of models which are more flexible and realistic. Hahn's approach allows for more complex applications than the discrete or the continuous models. A consumer might have income from work at unequal time intervals and/or make expenditures at unequal time intervals. Therefore, it is possible to obtain more rigorous and more accurate solutions with the approach here proposed.

In the first example we discuss the application of the Hahn quantum variational calculus to the Ramsey model, which determines the behavior of saving/consumption as the result of optimal inter-temporal choices by individual households (Atici and McMahan 2009). For a complete treatment of the classical Ramsey model we refer the reader to Barro and Sala-i-Martin (1999).

Example 2.61 Before writing the quantum model in terms of the Hahn operators we will present its discrete and continuous versions. The discrete-time Ramsey model is

$$\max_{[W_t]} \sum_{t=0}^{T-1} (1+p)^{-t} U\left[W_t - \frac{W_{t+1}}{1+r}\right], \quad C_t = W_t - \frac{W_{t+1}}{1+r},$$

while the continuous Ramsey model is

$$\max_{W(\cdot)} \int_0^T e^{-pt} U\left[rW(t) - W'(t)\right] dt, \quad C(t) = rW(t) - W'(t), \qquad (2.60)$$

where the quantities are defined as

- W – production function,
- C – consumption,
- p – discount rate,
- r – rate of yield,
- U – instantaneous utility function.

One may assume, due to some constraints of economical nature, that the dynamics do not depend on the usual derivative or the forward difference operator, but on the

Hahn quantum difference operator $D_{q,\omega}$. In this condition, one is entitled to assume again that the constraint $C(t)$ has the form

$$C(t) = -\left[E\left(-r, \frac{t-\omega}{q}\right)\right]^{-1} D_{q,\omega}\left[E\left(-r, \frac{t-\omega}{q}\right)W(t)\right],$$

where $E\left(z, \cdot\right)$ is the q, ω-exponential function defined by

$$E\left(z, t\right) := \prod_{k=0}^{\infty}(1 + zq^k(t(1-q) - \omega))$$

for $z \in \mathbb{C}$. Several nice properties of the q, ω-exponential function can be found in Aldwoah (2009); Annaby et al. (2012). By taking the q, ω-derivative of $\left[E\left(-r, \frac{t-\omega}{q}\right)W(t)\right]$ the following is obtained:

$$C(t) = -\left[E\left(-r, \frac{t-\omega}{q}\right)\right]^{-1}\left[E\left(-r, \frac{t-\omega}{q}\right)D_{q,\omega}W(t)\right.$$

$$\left. +E\left(-r, \frac{t-\omega}{q}\right)W(qt+\omega)\frac{r\left(1-\frac{1}{q}\right)-r\left(1+r\left(t-\frac{t-\omega}{q}\right)\right)}{\left(1+r\left(t-\frac{t-\omega}{q}\right)\right)(1-r(t(1-q)-\omega))}\right].$$

The quantum Ramsey model with the Hahn difference operator consists to maximize

$$\int_0^T E(-p,t)U\left[W(qt+\omega)\frac{r\left(1+r\left(t-\frac{t-\omega}{q}\right)\right)-r\left(1-\frac{1}{q}\right)}{\left(1+r\left(t-\frac{t-\omega}{q}\right)\right)(1-r(t(1-q)-\omega))}\right.$$

$$\left. - D_{q,\omega}W(t)\right]d_{q,\omega} \qquad (2.61)$$

subject to the constraint

$$C(t) = W(qt+\omega)\frac{r\left(1+r\left(t-\frac{t-\omega}{q}\right)\right)-r\left(1-\frac{1}{q}\right)}{\left(1+r\left(t-\frac{t-\omega}{q}\right)\right)(1-r(t(1-q)-\omega))} - D_{q,\omega}W(t). \quad (2.62)$$

The quantum Euler–Lagrange equation is, by Theorem 2.21, given by

$$E(-p,t)U'[C(t)]\frac{r\left(1+r\left(t-\frac{t-\omega}{q}\right)\right)-r\left(1-\frac{1}{q}\right)}{\left(1+r\left(t-\frac{t-\omega}{q}\right)\right)(1-r(t(1-q)-\omega))}$$

$$+ D_{q,\omega}\left[E(-p,t)U'[C(t)]\right] = 0. \qquad (2.63)$$

Note that for $q \uparrow 1$ and $\omega \downarrow 0$ problem (2.61)–(2.62) reduces to (2.60), and (2.63) to the classical Ramsey's Euler–Lagrange differential equation.

In the next example we analyze an adjustment model in economics. For a deeper discussion of this model we refer the reader to Sengupta (1997).

Example 2.62 Consider the dynamic model of adjustment

$$\mathcal{J}[y] = \sum_{t=1}^{T} r^t \left[\alpha(y(t) - \bar{y}(t))^2 + (y(t) - y(t-1))^2) \right] \longrightarrow \min,$$

where $y(t)$ is the output (state) variable, $r > 1$ is the exogenous rate of discount and $\bar{y}(t)$ is the desired target level, and T is the horizon. The first component of the loss function above is the disequilibrium cost due to deviations from desired target and the second component characterizes the agent's aversion to output fluctuations. In the continuous case the objective function has the form

$$\mathcal{J}[y] = \int_{1}^{T} e^{(r-1)t} \left[\alpha(y(t) - \bar{y}(t))^2 + (y'(t))^2 \right] \longrightarrow \min.$$

Let $q \in {]}0, 1{[}$ and $\omega \geq 0$ be fixed real numbers, and I be an interval of $\mathbb{R}$ such that $\omega_0, 0, T \in I$. The quantum model in terms of the Hahn operators which we wish to minimize is

$$\mathcal{J}[y] = \int_{0}^{T} E(1-r, t) \left[\alpha(y(qt+\omega) - \bar{y}(qt+\omega))^2 + (D_{q,\omega}[y](t))^2 \right] d_{q,\omega} t,$$
(2.64)

where $E(z, \cdot)$ is the q, ω-exponential function. By Theorem 2.51, a minimizer to (2.64) should satisfy the conditions

$$E(1-r, t) \left[\alpha(y(qt+\omega) - \bar{y}(qt+\omega)) \right] = D_{q,\omega} \left[E(1-r, t) D_{q,\omega}[y](t) \right], \quad (2.65)$$

for all $t \in \{\omega[n]_q : n \in \mathbb{N}_0\} \cup \{Tq^n + \omega[n]_q : n \in \mathbb{N}_0\} \cup \{\omega_0\}$; and

$$E(1-r, t) D_{q,\omega}[y](t)\big|_{t=0} = 0, \quad E(1-r, t) D_{q,\omega}[y](t)\big|_{t=T} = 0. \quad (2.66)$$

Taking the q, ω-derivative of the right side of (2.65) and applying properties of the q, ω-exponential function, for t such that $|t - \omega_0| < \frac{1}{(r-1)(1-q)}$, we can rewrite (2.65) and (2.66) as

$$[1 - (r-1)(t(1-q) - \omega)] \alpha(y(qt+\omega) - \bar{y}(qt+\omega))$$
$$= (r-1) D_{q,\omega}[y](t) + D_{q,\omega} D_{q,\omega}[y](t),$$
(2.67)

$$D_{q,\omega}[y](t)\big|_{t=0} = 0, \quad D_{q,\omega}[y](t)\big|_{t=T} = 0. \quad (2.68)$$

Note that for $(q, \omega) \to (1, 0)$ equations (2.67) and (2.68) reduce to

$$\alpha(y(t) - \bar{y}(t)) = (r - 1)y'(t) + y''(t),$$

$$y'(t)\big|_{t=0} = 0, \quad y'(t)\big|_{t=T} = 0,$$

which are necessary optimality conditions for the continuous model.

References

Aldwoah KA (2009) Generalized time scales and associated difference equations. Ph.D. Thesis, Cairo University, Cairo

Aldwoah KA, Hamza AE (2011) Difference time scales. Int J Math Stat 9(A11):106–125

Almeida R, Torres DFM (2009a) Hölderian variational problems subject to integral constraints. J Math Anal Appl 359(2):674–681

Almeida R, Torres DFM (2009b) Isoperimetric problems on time scales with nabla derivatives. J Vib Control 15(6):951–958

Almeida R, Torres DFM (2010a) Generalized Euler-Lagrange equations for variational problems with scale derivatives. Lett Math Phys 92(3):221–229

Almeida R, Torres DFM (2010b) Leitmann's direct method for fractional optimization problems. Appl Math Comput 217(3):956–962

Annaby MH, Hamza AE, Aldwoah KA (2012) Hahn difference operator and associated Jackson-Nörlund integrals. J Optim Theory Appl 154(1):133–153

Arutyunov AV (2000) Optimality conditions—abnormal and degenerate problems. Kluwer Academic Publishers, Dordrecht

Atici FM, McMahan CS (2009) A comparison in the theory of calculus of variations on time scales with an application to the Ramsey model. Nonlinear Dyn Syst Theory 9(1):1–10

Barro RJ, Sala-i-Martin X (1999) Economic growth. MIT Press, Cambridge

Caputo MR (2005) Foundations of dynamic economic analysis: optimal control theory and applications. Cambridge University Press, Cambridge

Carlson DA (2002) An observation on two methods of obtaining solutions to variational problems. J Optim Theory Appl 114(2):345–361

Carlson DA, Leitmann G (2005a) Coordinate transformation method for the extremization of multiple integrals. J Optim Theory Appl 127(3):523–533

Carlson DA, Leitmann G (2005b) A direct method for open-loop dynamic games for affine control systems. Dynamic games: theory and applications. Springer, New York, pp 37–55

Carlson DA, Leitmann G (2008) Fields of extremals and sufficient conditions for the simplest problem of the calculus of variations. J Global Optim 40(1–3):41–50

Cresson J, Frederico GSF, Torres DFM (2009) Constants of motion for non-differentiable quantum variational problems. Topol Methods Nonlinear Anal 33(2):217–231

Cruz PAF, Torres DFM, Zinober ASI (2010) A non-classical class of variational problems. Int J Math Model Numer Optim 1(3):227–236

Ernst T (2008) The different tongues of q-calculus. Proc Est Acad Sci 57(2):81–99

Ferreira RAC, Torres DFM (2008) Higher-order calculus of variations on time scales. Mathematical control theory and finance. Springer, Berlin, pp 149–159

Gouveia PDF, Torres DFM (2005) Automatic computation of conservation laws in the calculus of variations and optimal control. Comput Methods Appl Math 5(4):387–409

Gouveia PDF, Torres DFM, Rocha EAM (2006) Symbolic computation of variational symmetries in optimal control. Control Cybernet 35(4):831–849

Kac V, Cheung P (2002) Quantum calculus. Springer, New York

Koornwinder TH (1994) Compact quantum groups and q-special functions. Representations of Lie groups and quantum groups (Trento, 1993). Longman Sci Tech, Harlow, pp 46–128

Leitmann G (1967) A note on absolute extrema of certain integrals. Int J Non-Linear Mech 2:55–59

Leitmann G (2001a) On a class of direct optimization problems. J Optim Theory Appl 108(3):467–481

Leitmann G (2001b) Some extensions to a direct optimization method. J Optim Theory Appl 111(1):1–6

Leitmann G (2002) On a method of direct optimization. Vychisl Tekhnol 7:63–67

Leitmann G (2003) A direct method of optimization and its application to a class of differential games. Cubo Mat Educ 5(3):219–228

Leitmann G (2004) A direct method of optimization and its application to a class of differential games. Dyn Contin Discrete Impuls Syst Ser A Math Anal 11(2–3):191–204

Malinowska AB, Torres DFM (2009) Strong minimizers of the calculus of variations on time scales and the Weierstrass condition. Proc Est Acad Sci 58(4):205–212

Malinowska AB, Torres DFM (2010a) Leitmann's direct method of optimization for absolute extrema of certain problems of the calculus of variations on time scales. Appl Math Comput 217(3):1158–1162

Malinowska AB, Torres DFM (2010b) Natural boundary conditions in the calculus of variations. Math Methods Appl Sci 33(14):1712–1722

Malinowska AB, Torres DFM (2010c) The Hahn quantum variational calculus. J Optim Theory Appl 147(3):419–442

Martins N, Torres DFM (2009) Calculus of variations on time scales with nabla derivatives. Nonlinear Anal 71(12):e763–e773

Sengupta JK (1997) Recent models in dynamic economics: problems of estimating terminal conditions. Int J Syst Sci 28:857–864

Silva CJ, Torres DFM (2006) Absolute extrema of invariant optimal control problems. Commun Appl Anal 10(4):503–515

Torres DFM (2002) On the Noether theorem for optimal control. Eur J Control 8(1):56–63

Torres DFM (2004a) Proper extensions of Noether's symmetry theorem for nonsmooth extremals of the calculus of variations. Commun Pure Appl Anal 3(3):491–500

Torres DFM (2004b) Carathéodory equivalence Noether theorems, and Tonelli full-regularity in the calculus of variations and optimal control. J Math Sci (N Y) 120(1):1032–1050

Torres DFM, Leitmann G (2008) Contrasting two transformation-based methods for obtaining absolute extrema. J Optim Theory Appl 137(1):53–59

van Brunt B (2004) The calculus of variations. Springer, New York

Wagener FOO (2009) On the Leitmann equivalent problem approach. J Optim Theory Appl 142(1):229–242

Chapter 3
The Power Quantum Calculus

In this chapter we introduce the power difference calculus based on the operator $D_{n,q}[f](t) = \frac{f(qt^n)-f(t)}{qt^n-t}$, where n is an odd positive integer and $0 < q < 1$ (Aldwoah et al. 2012). Properties of the new operator and its inverse—the $d_{n,q}$ integral—are proved. As an application, we consider power quantum Lagrangian systems and corresponding n, q-Euler–Lagrange equations.

3.1 The Power Quantum Calculus

For a fixed $0 < q < 1$, $k \in \mathbb{N}_0 := \mathbb{N} \cup \{0\}$, and a fixed odd positive integer n, let us denote

$$\theta := \begin{cases} \infty & \text{if } n = 1, \\ q^{\frac{1}{1-n}} & \text{if } n \in 2\mathbb{N} + 1, \end{cases} \qquad S := \begin{cases} \{0\} & \text{if } n = 1, \\ \{-\theta, 0, \theta\} & \text{if } n \in 2\mathbb{N} + 1, \end{cases}$$

$$\text{and} \quad [k]_n := \begin{cases} \sum_{i=0}^{k-1} n^i & \text{if } k \in \mathbb{N}, \\ 0 & \text{if } k = 0. \end{cases}$$

Lemma 3.1 *Let $h : \mathbb{R} \longrightarrow \mathbb{R}$ be the function defined by $h(t) := qt^n$. Then, h is one-to-one, onto, and $h^{-1}(t) = \sqrt[n]{\frac{t}{q}}$. Moreover,*

$$h^k(t) := \underbrace{h \circ h \circ \cdots \circ h}_{k-times}(t) = q^{[k]_n} t^{n^k}$$

and

$$h^{-k}(t) := \underbrace{h^{-1} \circ h^{-1} \circ \cdots \circ h^{-1}}_{k-times}(t) = q^{-n^{-k}[k]_n} t^{n^{-k}}$$

A. B. Malinowska and D. F. M. Torres, *Quantum Variational Calculus*, SpringerBriefs in Control, Automation and Robotics, DOI: 10.1007/978-3-319-02747-0_3, © The Author(s) 2014

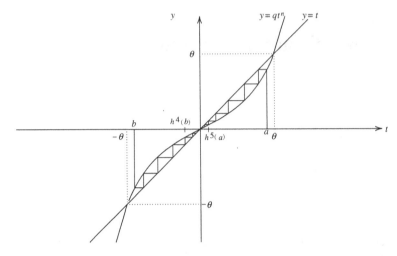

Fig. 3.1 The iteration of $h(t) = qt^n$, $t \in \mathbb{R}$, $n \in 2\mathbb{N} + 1$, $0 < q < 1$

with

$$\lim_{k \longrightarrow \infty} h^k(t) = \begin{cases} \infty & \text{if} \quad t > \theta \\ 0 & \text{if} -\theta < t < \theta \\ -\infty & \text{if} \quad t < -\theta \\ t & \text{if} \quad t \in S \end{cases}$$

and

$$\lim_{k \longrightarrow \infty} h^{-k}(t) = \begin{cases} \theta & \text{if} \quad 0 < t \\ -\theta & \text{if} \quad t < 0 \\ t & \text{if} \quad t \in S. \end{cases}$$

In Fig. 3.1 we illustrate the behaviour of $h^k(t)$ of Lemma 3.1 in the case $-\theta < t < \theta$.

3.1.1 Power Quantum Differentiation

We introduce the n,q-power difference operator as follows:

Definition 3.2 Assume that f is a real function defined on $\mathbb{R}$. The n,q-power operator is given by

$$D_{n,q}[f](t) := \begin{cases} \dfrac{f(qt^n) - f(t)}{qt^n - t} & \text{if } t \in \mathbb{R} \setminus S, \\ f'(t) & \text{if } t \in S, \end{cases}$$

provided f is differentiable at $t \in S$. If $D_{n,q}[f](t)$ exists, we say that f is n,q-differentiable at t.

The following lemma is a direct consequence of Definition 3.2.

Lemma 3.3 *Let f be a real function and $t \in \mathbb{R}$.*

(i) *If f is n,q-differentiable at t, $t \in S$, then f is continuous at t.*

(ii) *If f is n,q-differentiable on an interval $I \subset [-\theta, \theta]$, $0 \in I$, and*

$$D_{n,q}[f](t) = 0 \text{ for } t \in I,$$

then f is a constant function on I.

(iii) *If f is n,q-differentiable at t, then $f(qt^n) = f(t) + (qt^n - t)D_{n,q}[f](t)$.*

The next theorem gives useful formulas for the computation of n,q-derivatives of sums, products, and quotients of n,q-differentiable functions.

Theorem 3.4 *Assume $f, g : \mathbb{R} \longrightarrow \mathbb{R}$ are n,q-differentiable at $t \in \mathbb{R}$. Then:*

(i) *The sum $f + g : \mathbb{R} \longrightarrow \mathbb{R}$ is n,q-differentiable at t and*

$$D_{n,q}[f + g](t) = D_{n,q}[f](t) + D_{n,q}[g](t).$$

(ii) *For any constant c, $cf : \mathbb{R} \longrightarrow \mathbb{R}$ is n,q-differentiable at t and*

$$D_{n,q}[cf](t) = cD_{n,q}[f](t).$$

(iii) *The product $fg : \mathbb{R} \longrightarrow \mathbb{R}$ is n,q-differentiable at t and*

$$\begin{aligned} D_{n,q}[fg](t) &= D_{n,q}[f](t)g(t) + f(qt^n)D_{n,q}[g](t) \\ &= f(t)D_{n,q}[g](t) + D_{n,q}[f](t)g(qt^n). \end{aligned}$$

(iv) *If $g(t)g(qt^n) \neq 0$, then f/g is n,q-differentiable at t and*

$$D_{n,q}\left[\frac{f}{g}\right](t) = \frac{D_{n,q}[f](t)g(t) - f(t)D_{n,q}[g](t)}{g(t)g(qt^n)}.$$

Proof The proof is done by direct calculations.

Next example gives explicit formulas for the n,q-derivative of some simple functions.

Example 3.5 Let $f : \mathbb{R} \longrightarrow \mathbb{R}$.

(i) If $f(t) = c$ for all $t \in \mathbb{R}$, where $c \in \mathbb{R}$ is a constant, then $D_{n,q}[f](t) = 0$.

(ii) If $f(t) = t$ for all $t \in \mathbb{R}$, then $D_{n,q}[f](t) = 1$.

(iii) If $f(t) = at - b$ for all $t \in \mathbb{R}$, where a, b are real constants, then by Theorem 3.4 we have $D_{n,q}[f](t) = a$.

(iv) If $f(t) = t^2$ for all $t \in \mathbb{R}$, then $D_{n,q}[f](t) = t + qt^n$.

(v) If $f(t) = \frac{1}{t}$ for all $t \in \mathbb{R} \setminus \{0\}$, then $D_{n,q}[f](t) = -\frac{1}{qt^{n+1}}$.

(vi) If $f(t) = (t + b)^m$ for all $t \in \mathbb{R}$, where $b \in \mathbb{R}$ is a constant and $m \in \mathbb{N}$, then, by induction on m, we obtain that

$$D_{n,q}[f](t) = \sum_{k=0}^{m-1} (qt^n + b)^k (t + b)^{m-1-k}$$

for $t \neq S$.

We note that by definition of the n,q-difference operator, one has

$$D_{n,q}[f](t) = f'(t), \quad t \in S,$$

for all functions f in (i)–(vi).

Definition 3.6 Let $f : \mathbb{R} \longrightarrow \mathbb{R}$. We define the second n,q-derivative by $D_{n,q}^2[f] := D_{n,q}[D_{n,q}[f]]$. More generally, we define $D_{n,q}^m[f]$ as follows:

$$D_{n,q}^0[f] = f,$$
$$D_{n,q}^m[f] = D_{n,q}[D_{n,q}^{m-1}[f]], \quad m \in \mathbb{N}.$$

We now obtain, under certain conditions, the formula for the mth n,q-derivative of fg, $m \in \mathbb{N}$.

Let h be the function defined in Lemma 3.1 and let us write $h^\circ f$ to denote $f \circ h$. We will denote by S_k^m the set consisting of all possible strings of length m, containing exactly k times h° and $m - k$ times $D_{n,q}$.

Example 3.7 Let $k = 2$ and $m = 4$. Then,

$$S_2^4 = \Big\{ D_{n,q} D_{n,q} h^\circ h^\circ, \ D_{n,q} h^\circ D_{n,q} h^\circ, \ D_{n,q} h^\circ h^\circ D_{n,q}, \ h^\circ h^\circ D_{n,q} D_{n,q},$$
$$h^\circ D_{n,q} h^\circ D_{n,q}, \ h^\circ D_{n,q} D_{n,q} h^\circ \Big\}.$$

Example 3.8 If $m = 2$, then for $k = 0, 1, 2$ we have

$$S_0^2 = \Big\{ D_{n,q} D_{n,q} \Big\}, \quad S_1^2 = \Big\{ D_{n,q} h^\circ, h^\circ D_{n,q} \Big\}, \quad S_2^2 = \Big\{ h^\circ h^\circ \Big\}.$$

Let $f : \mathbb{R} \longrightarrow \mathbb{R}$. Then,

$$(D_{n,q} h^\circ f)(t) = D_{n,q}[f \circ h](t),$$
$$(h^\circ D_{n,q} f)(t) = D_{n,q}[f](h(t)),$$

provided all these quantities exist. Observe that $D_{n,q}[f \circ h](t) \neq D_{n,q}[f](h(t))$.
Indeed,

$$
\begin{aligned}
D_{n,q}[f](h(t)) &= \frac{f(h(h(t))) - f(h(t))}{h(h(t)) - h(t)} \\
&= \frac{f(h(h(t))) - f(h(t))}{(h(t) - t)D_{n,q}[h](t)} = \frac{D_{n,q}[f \circ h](t)}{D_{n,q}[h](t)}.
\end{aligned}
$$

Theorem 3.9 **(Leibniz formula)** *Let $\mathcal{S}_k^m$ be the set consisting of all possible strings of length m, containing exactly k times h° and $m - k$ times $D_{n,q}$. If f is a function for which Lf exist for all $L \in \mathcal{S}_k^m$, and function g is m times n,q-differentiable, then for all $m \in \mathbb{N}$ we have:*

$$
D_{n,q}^m[fg](t) = \sum_{k=0}^{m} \left(\sum_{L \in \mathcal{S}_k^m} Lf \right)(t) \, D_{n,q}^k[g](t) \quad for \ t \in \mathbb{R} \setminus S, \tag{3.1}
$$

and

$$
D_{n,q}^m[fg](t) = \sum_{k=0}^{m} \binom{m}{k} D_{n,q}^{m-k}[f](t) \, D_{n,q}^k[g](t) \quad for \ t \in S. \tag{3.2}
$$

Proof For $t \in S$ equality (3.2) yields

$$
(fg)^{(m)}(t) = \sum_{k=0}^{m} \binom{m}{k} f^{(m-k)}(t) \, g^{(k)}(t),
$$

which is true (the standard Leibniz formula of classical calculus). Assume $t \notin S$.
The proof is done by induction on m. If $m = 1$, then by Theorem 3.4 we have
$D_{n,q}[fg](t) = D_{n,q}[f](t)g(t) + h^\circ f(t)D_{n,q}[g](t)$, i.e., (3.1) is true for $m = 1$. We
now assume that (3.1) is true for $m = s$ and prove that it is also true for $m = s + 1$.
First, we note that for $k \in \mathbb{N}$ and $t \notin S$

$$
\begin{aligned}
D_{n,q}^{m+1}[fg](t) &= D_{n,q} \left[\sum_{k=0}^{m} \left(\sum_{L \in \mathcal{S}_k^m} Lf \right)(t) \, D_{n,q}^k[g](t) \right] \\
&= \sum_{k=0}^{m} \left[D_{n,q} \left(\sum_{L \in \mathcal{S}_k^m} Lf \right)(t) \, D_{n,q}^k[g](t) + h^\circ \left(\sum_{L \in \mathcal{S}_k^m} Lf \right)(t) \, D_{n,q}^{k+1}[g](t) \right] \\
&= \sum_{k=0}^{m} \left(\sum_{L \in \mathcal{S}_k^m} D_{n,q}Lf \right)(t) \, D_{n,q}^k[g](t) + \sum_{k=1}^{m+1} \left(\sum_{L \in \mathcal{S}_{k-1}^m} h^\circ Lf \right)(t) \, D_{n,q}^k[g](t)
\end{aligned}
$$

$$= \left(\sum_{L \in \mathcal{S}_m^m} h \circ Lf \right)(t)\, D_{n,q}^{m+1}[g](t) + \left(\sum_{L \in \mathcal{S}_0^m} D_{n,q} Lf \right)(t)\, g(t)$$

$$+ \sum_{k=1}^{m} \left(\sum_{L \in \mathcal{S}_{k-1}^m} h \circ L\, f(t) + \sum_{L \in \mathcal{S}_k^m} D_{n,q} L\, f \right)(t)\, D_{n,q}^k[g](t)$$

$$= \left(\sum_{L \in \mathcal{S}_{m+1}^{m+1}} Lf \right)(t)\, D_{n,q}^{m+1}[g](t) + \left(\sum_{L \in \mathcal{S}_0^{m+1}} Lf \right)(t)\, g(t)$$

$$+ \sum_{k=1}^{m} \left(\sum_{L \in \mathcal{S}_k^{m+1}} Lf \right)(t)\, D_{n,q}^k[g](t)$$

$$= \sum_{k=0}^{m+1} \left(\sum_{L \in \mathcal{S}_k^{m+1}} Lf \right)(t)\, D_{n,q}^k[g](t).$$

We conclude that (3.1) is true for $m = s + 1$. Hence, by mathematical induction, (3.1) holds for all $m \in \mathbb{N}$ and $t \in \mathbb{R} \setminus S$.

The standard chain rule of classical calculus does not necessarily hold true for the n,q-quantum calculus. For example, if we assume that $f, g : \mathbb{R} \longrightarrow \mathbb{R}$ are defined by $f(t) = t^2$ and $g(t) = qt$, then we have

$$D_{n,q}[f \circ g](t) = D_{n,q}(qt)^2 = D_{n,q}(q^2 t^2) = q^2(t + qt^n)$$
$$\neq q^2(t + q^n t^n) = D_{n,q}[f](g(t)) \cdot D_{n,q}[g](t).$$

However, we can derive an analogous formula of the chain rule for our power quantum calculus.

Theorem 3.10 (Power chain rule) *Assume* $g : I \longrightarrow \mathbb{R}$ *is continuous and* n,q-*differentiable, and* $f : \mathbb{R} \longrightarrow \mathbb{R}$ *is continuously differentiable. Then there exists a constant* c *between* qt^n *and* t *with*

$$D_{n,q}[f \circ g](t) = f'(g(c)) D_{n,q}[g](t). \tag{3.3}$$

Proof For $t \notin S$ we have

$$D_{n,q}[f \circ g](t) = \frac{f(g(qt^n)) - f(g(t))}{qt^n - t}.$$

We may assume that $g(qt^n) \neq g(t)$ (because if $g(qt^n) = g(t)$, then $D_{n,q}[f \circ g](t) = D_{n,q}[g](t) = 0$ and (3.3) holds for any c between qt^n and t). Then,

$$D_{n,q}[f \circ g](t) = \frac{f(g(qt^n)) - f(g(t))}{g(qt^n) - g(t)} \cdot \frac{g(qt^n) - g(t)}{qt^n - t}. \tag{3.4}$$

By the mean value theorem, there exists a real number τ between $g(t)$ and $g(qt^n)$ with

$$\frac{f(g(qt^n)) - f(g(t))}{g(qt^n) - g(t)} = f'(\tau). \tag{3.5}$$

In view of the continuity of g, there exists c in the interval with end points qt^n and t such that $g(c) = \tau$. Thus from (3.4) and (3.5) we obtain (3.3). Relation (3.3) is true at t, $t \in S$, by the classical chain rule.

3.1.2 Power Quantum Integration

In this section we are interested to study the inverse operation of $D_{n,q}[f]$. We call this inverse the n,q-integral of f (or the power quantum integral). We define the interval I to be $[-\theta, \theta]$.

Definition 3.11 Let $f : I \longrightarrow \mathbb{R}$ and $a, b \in I$. We say that F is a n,q-antiderivative of f on I if $D_{n,q}[F](t) = f(t)$ for all $t \in I$.

From now on we assume that all series considered along the text are convergent.

Theorem 3.12 Let $f : I \longrightarrow \mathbb{R}$ and $a, b \in I$. The function

$$F(t) = -\sum_{k=0}^{\infty} q^{[k]_n} t^{n^k} \left(q^{n^k} t^{n^k(n-1)} - 1 \right) f\left(q^{[k]_n} t^{n^k} \right)$$

is a n,q-antiderivative of f on I, provided f is continuous at 0.

Proof For $t \neq 0$, we have

$$D_{n,q}[F](t) = \frac{F(qt^n) - F(t)}{qt^n - t}$$

$$= \sum_{k=0}^{\infty} \left[-\frac{q^{[k+1]_n} t^{n^{k+1}}}{qt^n - t} \left(q^{n^{k+1}} t^{n^{k+1}(n-1)} - 1 \right) f\left(q^{[k+1]_n} t^{n^{k+1}} \right) \right.$$

$$\left. + \frac{q^{[k]_n} t^{n^k} \left(q^{n^k} t^{n^k(n-1)} - 1 \right)}{qt^n - t} f\left(q^{[k]_n} t^{n^k} \right) \right]$$

$$= f(t).$$

If $t = 0$, then the continuity of f at 0 implies that

$$D_{n,q}[F](0) = \lim_{s \to 0} \frac{F(s) - F(0)}{s}$$

$$= \lim_{s \to 0} \frac{-\sum_{k=0}^{\infty} q^{[k]_n s^{n^k}} \left(q^{n^k} s^{n^k(n-1)} - 1 \right) f \left(q^{[k]_n} s^{n^k} \right)}{s}$$

$$= \lim_{s \to 0} -\sum_{k=0}^{\infty} q^{[k]_n s^{n^k}-1} \left(q^{n^k} s^{n^k(n-1)} - 1 \right) f \left(q^{[k]_n} s^{n^k} \right)$$

$$= \lim_{s \to 0} -\sum_{k=0}^{\infty} \left(q^{[k+1]_n s^{n^k}+n^k(n-1)-1} - q^{[k]_n s^{n^k}-1} \right) f \left(q^{[k]_n} s^{n^k} \right)$$

$$= \lim_{s \to 0} -\sum_{k=0}^{\infty} \left(q^{[k+1]_n s^{n^{k+1}}-1} - q^{[k]_n s^{n^k}-1} \right) f \left(q^{[k]_n} s^{n^k} \right)$$

$$= \lim_{s \to 0} f(s)$$

$$= f(0).$$

This completes the proof.

We then define the indefinite n,q-integral of f by

$$\int_I f(t) \, d_{n,q}t := F(t) + C,$$

where C is an arbitrary constant. The definite n,q-integral of f is defined as follows.

Definition 3.13 Let $f : I \longrightarrow \mathbb{R}$ and $a, b \in I$. We define the n,q-integral of f from a to b by

$$\int_a^b f(t) \, d_{n,q}t := \int_0^b f(t) \, d_{n,q}t - \int_0^a f(t) \, d_{n,q}t, \tag{3.6}$$

where

$$\int_0^x f(t) \, d_{n,q}t := -\sum_{k=0}^{\infty} q^{[k]_n x^{n^k}} \left(q^{n^k} x^{n^k(n-1)} - 1 \right) f \left(q^{[k]_n} x^{n^k} \right), \quad x \in I, \tag{3.7}$$

provided the series at the right-hand side of (3.7) converge at $x = a$ and $x = b$.

Definition 3.14 A function f is said to be n,q-integrable on a subinterval J of I if

$$\left| \int_a^b f(t)\, d_{n,q}t \right| < \infty \quad \text{for all } a, b \in J.$$

Remark 3.15 The integral formulas (3.6) and (3.7) yield

$$\int_a^b f(t)d_qt = \int_0^b f(t)d_qt - \int_0^a f(t)d_qt$$

and

$$\int_0^a f(t)d_qt = a(1 - q) \sum_{k=0}^{\infty} q^k f(aq^k)$$

when $n = 1$; and yield the corresponding integral of the operator D_n defined by

$$D_n f(t) = \begin{cases} \frac{f(t^n) - f(t)}{t^n - t} & \text{if } t \in \mathbb{R} \setminus \{-1, 0, 1\}, \\ f'(t) & \text{if } t \in \{-1, 0, 1\}, \end{cases}$$

when $q \to 1$.

The following properties of the n,q-integral are direct consequences of the definition and provide extensions of analogous properties of the Jackson q-integral (Jackson 1908, 1910; Kac and Cheung 2002).

Lemma 3.16 *Let* $f, g : I \longrightarrow \mathbb{R}$ *be* n,q-*integrable,* $k \in \mathbb{R}$, *and* $a, b, c \in I$. *Then,*

(i) $\displaystyle\int_a^a f(t)\, d_{n,q}t = 0.$

(ii) $\displaystyle\int_a^b k f(t)\, d_{n,q}t = k \int_a^b f(t)\, d_{n,q}t.$

(iii) $\displaystyle\int_a^b f(t)\, d_{n,q}t = - \int_b^a f(t)\, d_{n,q}t.$

(iv) $\displaystyle\int_a^b f(t)\, d_{n,q}t = \int_a^c f(t)\, d_{n,q}t + \int_c^b f(t)\, d_{n,q}t \text{ for } a \leq c \leq b.$

(v) $\displaystyle\int_a^b (f(t) + g(t))\, d_{n,q}t = \int_a^b f(t)\, d_{n,q}t + \int_a^b g(t)\, d_{n,q}t.$

Theorem 3.17 *Assume that* $f : I \longrightarrow \mathbb{R}$ *is continuous at* 0. *Then,*

$$\int_a^b D_{n,q} f(t) d_{n,q} t = f(b) - f(a) \quad \text{for all } a, b \in I.$$

Proof First, we note that $\lim_{r \to \infty} q^{[r]_n a^{n^r}} = \lim_{r \to \infty} q^{[r]_n b^{n^r}} = 0$. By the continuity of f at 0,

$$\lim_{r \to 0} f(r) = \lim_{k \to \infty} f(q^{[k]_n a^{n^k}}) = \lim_{k \to \infty} f(q^{[k]_n b^{n^k}}) = f(0).$$

Thus,

$$\int_a^b D_{n,q}[f](t) \, d_{n,q} t = - \sum_{k=0}^{\infty} q^{[k]_n b^{n^k}} \left(q^{n^k} b^{n^k(n-1)} - 1 \right) D_{n,q}[f] \left(q^{[k]_n b^{n^k}} \right)$$

$$+ \sum_{k=0}^{\infty} q^{[k]_n a^{n^k}} \left(q^{n^k} a^{n^k(n-1)} - 1 \right) D_{n,q}[f] \left(q^{[k]_n a^{n^k}} \right)$$

$$= \sum_{k=0}^{\infty} \left[-q^{[k]_n b^{n^k}} \left(q^{n^k} b^{n^k(n-1)} - 1 \right) \frac{f \left(q^{[k+1]_n b^{n^{k+1}}} \right) - f \left(q^{[k]_n b^{n^k}} \right)}{q^{[k+1]_n b^{n^{k+1}}} - q^{[k]_n b^{n^k}}} \right]$$

$$+ \sum_{k=0}^{\infty} \left[q^{[k]_n a^{n^k}} \left(q^{n^k} a^{n^k(n-1)} - 1 \right) \frac{f \left(q^{[k+1]_n a^{n^{k+1}}} \right) - f \left(q^{[k]_n a^{n^k}} \right)}{q^{[k+1]_n a^{n^{k+1}}} - q^{[k]_n a^{n^k}}} \right]$$

$$= f(b) - f(a).$$

This completes the proof.

Lemma 3.18 *Let* $s \in J \subseteq [0, \theta]$ *and* g *be* n, q-*integrable on* J. *If* $0 \le |f(t)| \le g(t)$ *for all* $t \in \left\{ q^{[k]_n s^{n^k}} : k \in \mathbb{N}_0 \right\}$, *then*

$$\left| \int_0^b f(t) d_{n,q} t \right| \le \int_0^b g(t) d_{n,q} t \quad \text{and} \quad \left| \int_a^b f(t) d_{n,q} t \right| \le \int_a^b g(t) d_{n,q} t \quad (3.8)$$

for $a, b \in \left\{ q^{[k]_n s^{n^k}} : k \in \mathbb{N}_0 \right\}$ *with* $a < b$. *Consequently, if* $g(t) \ge 0$ *for all* $t \in \left\{ q^{[k]_n s^{n^k}} : k \in \mathbb{N}_0 \right\}$, *then*

$$\int_0^b g(t)d_{n,q}t \geq 0 \quad and \quad \int_a^b g(t)d_{n,q}t \geq 0 \tag{3.9}$$

for all $a, b \in \left\{ q^{[k]_n} s^{n^k} : k \in \mathbb{N}_0 \right\}$ such that $a < b$.

Proof If $b \in \left\{ q^{[k]_n} s^{n^k} : k \in \mathbb{N}_0 \right\}$, then we can write $b = q^{[k_2]_n} s^{n^{k_2}}$ for some $k_2 \in \mathbb{N}_0$. Observe that, for all $k \in \mathbb{N}_0$,

$$q^{[k]_n} b^{n^k} = q^{[k+k_2]_n} s^{n^{k+k_2}} \in \left\{ q^{[k]_n} s^{n^k} : k \in \mathbb{N}_0 \right\}.$$

Therefore, by assumption, we have $0 \leq |f(q^{[k]_n} b^{n^k})| \leq g(q^{[k]_n} b^{n^k})$ and $-q^{[k]_n} b^{n^k} \left(q^{n^k} b^{n^k(n-1)} - 1 \right) > 0$ for all $k \in \mathbb{N}_0$. Since g is n,q-integrable on J, it follows that the series

$$\sum_{k=0}^{\infty} -q^{[k]_n} b^{n^k} \left(q^{n^k} b^{n^k(n-1)} - 1 \right) f \left(q^{[k]_n} b^{n^k} \right)$$

is absolutely convergent. Therefore,

$$\left| \int_0^b f(t)d_{n,q}t \right| = \left| \sum_{k=0}^{\infty} -q^{[k]_n} b^{n^k} \left(q^{n^k} b^{n^k(n-1)} - 1 \right) f \left(q^{[k]_n} b^{n^k} \right) \right|$$

$$\leq \sum_{k=0}^{\infty} -q^{[k]_n} b^{n^k} \left(q^{n^k} b^{n^k(n-1)} - 1 \right) \left| f \left(q^{[k]_n} b^{n^k} \right) \right|$$

$$\leq \sum_{k=0}^{\infty} -q^{[k]_n} b^{n^k} \left(q^{n^k} b^{n^k(n-1)} - 1 \right) g \left(q^{[k]_n} b^{n^k} \right)$$

$$= \int_0^b g(t)d_{n,q}t.$$

Now, if $a, b \in \left\{ q^{[k]_n} s^{n^k} : k \in \mathbb{N}_0 \right\}$ and $a < b$, then we can write $a = q^{[k_1]_n} s^{n^{k_1}}$ and $b = q^{[k_2]_n} s^{n^{k_2}}$ for some $k_1, k_2 \in \mathbb{N}, k_1 > k_2$. Hence,

$$\left| \int_a^b f(t)d_{n,q}t \right|$$

$$= \left| -\sum_{k=0}^{\infty} q^{[k+k_2]_n} s^{n^{k+k_2}} \left(q^{n^{k+k_2}} s^{n^{k+k_2}(n-1)} - 1 \right) f \left(q^{[k+k_2]_n} s^{n^{k+k_2}} \right) \right.$$

$$+ \sum_{k=0}^{\infty} q^{[k+k_1]_n} s^{n^{k+k_1}} \left(q^{n^{k+k_1}} s^{n^{k+k_1}(n-1)} - 1 \right) f \left(q^{[k+k_1]_n} s^{n^{k+k_1}} \right) \Bigg|$$

$$= \Bigg| \sum_{k=k_2}^{\infty} q^{[k]_n} s^{n^k} \left(1 - q^{n^k} s^{n^k(n-1)} \right) f \left(q^{[k]_n} s^{n^k} \right)$$

$$- \sum_{k=k_1}^{\infty} q^{[k]_n} s^{n^k} \left(1 - q^{n^k} s^{n^k(n-1)} \right) f \left(q^{[k]_n} s^{n^k} \right) \Bigg|$$

$$\leq \sum_{k=k_2}^{k_1-1} q^{[k]_n} s^{n^k} \left(1 - q^{n^k} s^{n^k(n-1)} \right) \left| f \left(q^{[k]_n} s^{n^k} \right) \right|$$

$$\leq \sum_{k=k_2}^{k_1-1} q^{[k]_n} s^{n^k} \left(1 - q^{n^k} s^{n^k(n-1)} \right) g \left(q^{[k]_n} s^{n^k} \right)$$

$$\mp \sum_{k=k_1}^{\infty} q^{[k]_n} s^{n^k} \left(1 - q^{n^k} s^{n^k(n-1)} \right) g \left(q^{[k]_n} s^{n^k} \right)$$

$$= \sum_{k=k_2}^{\infty} q^{[k]_n} s^{n^k} \left(1 - q^{n^k} s^{n^k(n-1)} \right) g \left(q^{[k]_n} s^{n^k} \right)$$

$$- \sum_{k=k_1}^{\infty} q^{[k]_n} s^{n^k} \left(1 - q^{n^k} s^{n^k(n-1)} \right) g \left(q^{[k]_n} s^{n^k} \right)$$

$$= - \sum_{k=0}^{\infty} q^{[k+k_2]_n} s^{n^{k+k_2}} \left(q^{n^{k+k_2}} s^{n^{k+k_2}(n-1)} - 1 \right) g \left(q^{[k+k_2]_n} s^{n^{k+k_2}} \right)$$

$$+ \sum_{k=0}^{\infty} q^{[k+k_1]_n} s^{n^{k+k_1}} \left(q^{n^{k+k_1}} s^{n^{k+k_1}(n-1)} - 1 \right) g \left(q^{[k+k_1]_n} s^{n^{k+k_1}} \right)$$

$$= \int_{a}^{b} g(t) d_{n,q} t.$$

To show that (3.9) is true, we just put $f = 0$ in (3.8).

Lemma 3.19 Let $s \in J \subseteq [-\theta, 0]$ and g be n,q-integrable on J. If $0 \leq |f(t)| \leq g(t)$ for all $t \in \left\{ q^{[k]_n} s^{n^k} : k \in \mathbb{N}_0 \right\}$, then

$$\left| \int_{b}^{0} f(t) d_{n,q} t \right| \leq \int_{b}^{0} g(t) d_{n,q} t \quad \text{and} \quad \left| \int_{a}^{b} f(t) d_{n,q} t \right| \leq \int_{a}^{b} g(t) d_{n,q} t$$

for $a, b \in \left\{ q^{[k]_n} s^{n^k} : k \in \mathbb{N}_0 \right\}$ such that $a < b$. Consequently, if $g(t) \geq 0$ for all $t \in \left\{ q^{[k]_n} s^{n^k} : k \in \mathbb{N}_0 \right\}$, then

$$\int\limits_b^0 g(t) d_{n,q} t \geq 0 \qquad and \qquad \int\limits_a^b g(t) d_{n,q} t \geq 0$$

for all $a, b \in \left\{ q^{[k]_n} s^{n^k} : k \in \mathbb{N}_0 \right\}$ such that $a < b$.

Proof Arguing as in the proof of Lemma 3.18, we can show that the series

$$\sum_{k=0}^{\infty} -q^{[k]_n} b^{n^k} \left(q^{n^k} b^{n^k(n-1)} - 1 \right) f \left(q^{[k]_n} b^{n^k} \right)$$

is absolutely convergent. Therefore, we have

$$\left| \int\limits_b^0 f(t) d_{n,q} t \right| = \left| \int\limits_0^b f(t) d_{n,q} t \right|$$

$$= \left| -\sum_{k=0}^{\infty} q^{[k]_n} b^{n^k} \left(q^{n^k} b^{n^k(n-1)} - 1 \right) f \left(q^{[k]_n} b^{n^k} \right) \right|$$

$$\leq \sum_{k=0}^{\infty} \left| -q^{[k]_n} b^{n^k} \left(q^{n^k} b^{n^k(n-1)} - 1 \right) \right| \left| f \left(q^{[k]_n} b^{n^k} \right) \right|$$

$$\leq \sum_{k=0}^{\infty} q^{[k]_n} b^{n^k} \left(q^{n^k} b^{n^k(n-1)} - 1 \right) g \left(q^{[k]_n} b^{n^k} \right)$$

$$= -\int\limits_0^b g(t) d_{n,q} t = \int\limits_b^0 g(t) d_{n,q} t.$$

The rest of the proof can be done similarly to the proof of Lemma 3.18.

It should be noted that the inequality

$$\left| \int\limits_a^b f(t) d_{n,q} t \right| \leq \int\limits_a^b |f(t)| d_{n,q} t \quad \text{for all } a, b \in I$$

is not always true. For example, fix $n = 1$, $0 < q < 1$, and define the function $f : [0, 1] \longrightarrow \mathbb{R}$ by

$$f(x) = \begin{cases} \frac{1}{1-q}\left(4q^{-n}x - (1+3q)\right), & q^{n+1} \le x \le \frac{q^n(1+q)}{2}, \ n \in \mathbb{N}, \\ \frac{4}{1-q}\left(-xq^{-n}+1\right) - 1, & \frac{q^n(1+q)}{2} \le x \le q^n, \quad n \in \mathbb{N}, \\ 0, & x = 0, \end{cases}$$

(see Abu Risha et al. 2007). It follows that f is n,q-integrable on $[0, 1]$, $f(q^n) = -1$, and $f\left(\frac{1+q}{2}q^n\right) = 1$ for all $n \in \mathbb{N}$. By a direct calculation one can see that

$$\int_{\frac{1+q}{2}}^{1} f(t) \, d_{n,q}t = -\frac{3+q}{2} \quad \text{and} \quad \int_{\frac{1+q}{2}}^{1} |f(t)| \, d_{n,q}t = \frac{1-q}{2}.$$

Thus,

$$\left| \int_{\frac{1+q}{2}}^{1} f(t) \, d_{n,q}t \right| > \int_{\frac{1+q}{2}}^{1} |f(t)| \, d_{n,q}t.$$

Lemma 3.20 *Let $f, g : I \longrightarrow \mathbb{R}$.*

(i) *If functions f and g are n,q-differentiable, then the following integration by parts formula holds:*

$$\int_{a}^{b} f(t) D_{n,q}[g](t) \, d_{n,q}t = f(t)g(t)\big|_{a}^{b} - \int_{a}^{b} D_{n,q}[f](t)g(qt^n) \, d_{n,q}t, \quad a, b \in I.$$

$$(3.10)$$

(ii) *If f is continuous at 0, then for $t \in I$*

$$\int_{t}^{qt^n} f(r) \, d_{n,q}r = (qt^n - t)f(t).$$

Proof (i) By Theorem 3.17 we have .

$$\int_{a}^{b} D_{n,q}[fg](t)d_{n,q}t = (fg)(b) - (fg)(a).$$

On the other hand, by (iii) of Theorem 3.4 and (v) of Theorem 3.16,

$$\int_{a}^{b} D_{n,q}[fg](t) = \int_{a}^{b} f(t) D_{n,q}[g](t)d_{n,q}t + \int_{a}^{b} D_{n,q}[f](t)g(qt^n)d_{n,q}t.$$

Combining these two equalities we get the desired formula.

(ii)

$$\int_t^{qt^n} f(s)\, d_{n,q}s = \int_0^{qt^n} f(s)\, d_{n,q}s - \int_0^t f(s)\, d_{n,q}s$$

$$= \sum_{k=0}^{\infty} \left[q^{[k]_n t^{n^k}} \left(q^{n^k} t^{n^k(n-1)} - 1 \right) f \left(q^{[k]_n} t^{n^k} \right) \right.$$

$$\left. - q^{[k+1]_n} t^{n^{k+1}} \left(q^{n^{k+1}} t^{n^{k+1}(n-1)} - 1 \right) f \left(q^{[k+1]_n} t^{n^{k+1}} \right) \right]$$

$$= (qt^n - t) f(t).$$

3.2 The Power Quantum Variational Calculus

In this section we give one application of the power quantum calculus which we derived in Sect. 3.1, introducing the power quantum variational calculus and proving a quantum analog of the Euler–Lagrange equation (Sect. 3.2.1). This provides a necessary optimality condition for local minimizers. Direct methods can also be developed for our power quantum variational calculus, allowing to obtain global minimizers for certain classes of problems (Sect. 3.2.2).

As in Sect. 3.1.2, we define the interval I to be $[-\theta, \theta]$. Let $a, b \in I$ with $a < b$. We define the n,q-interval by

$$[a, b]_{n,q} := \left\{ q^{[k]_n} a^{n^k} : k \in \mathbb{N}_0 \right\} \cup \left\{ q^{[k]_n} b^{n^k} : k \in \mathbb{N}_0 \right\} \cup \{0\}.$$

Let $\mathcal{D}([a, b]_{n,q}, \mathbb{R})$ be the set of all real valued functions continuous and bounded on $[a, b]_{n,q}$.

Lemma 3.21 (Fundamental Lemma of the power quantum variational calculus) *Let $f \in \mathcal{D}([a, b]_{n,q}, \mathbb{R})$. One has $\int_a^b f(t)g(qt^n)\, d_{n,q}t = 0$ for all functions $g \in \mathcal{D}([a, b]_{n,q}, \mathbb{R})$ with $g(a) = g(b) = 0$ if and only if $f(t) = 0$ for all $t \in [a, b]_{n,q}$.*

Proof The implication "$\Leftarrow$" is obvious. Let us prove the implication "$\Rightarrow$". Suppose, by contradiction, that $f(c) \neq 0$ for some $c \in [a, b]_{n,q}$.

Case I If $c \neq 0$, then without loss of generality we can assume that $c = q^{[k]_n} a^{n^k}$ for $k \in \mathbb{N}_0$. Define

$$g(t) = \begin{cases} f \left(q^{[k]_n} a^{n^k} \right) & \text{if } t = q^{[k+1]_n} a^{n^{k+1}} \\ 0 & \text{otherwise} \end{cases}$$

Since $a \neq 0$, we see that

$$\int_a^b f(t)g(qt^n)\,d_{n,q}t = q^{[k]_n}a^{n^k}\left(q^{n^k}a^{n^k(n-1)} - 1\right)\left(f\left(q^{[k]_n}a^{n^k}\right)\right)^2 \neq 0,$$

which is a contradiction.

Case II If $c = 0$, then without loss of generality we can assume that $f(0) > 0$. We know that (see Lemma 3.1)

$$\lim_{k\to\infty} q^{[k]_n}a^{n^k} = \lim_{k\to\infty} q^{[k]_n}b^{n^k} = 0.$$

As f is continuous at 0,

$$\lim_{k\to\infty} f\left(q^{[k]_n}a^{n^k}\right) = \lim_{k\to\infty} f\left(q^{[k]_n}b^{n^k}\right) = f(0).$$

Therefore, there exists $N \in \mathbb{N}$ such that for all $l > N$ the inequalities

$$f\left(q^{[l]_n}a^{n^l}\right) > 0, \quad f\left(q^{[l]_n}b^{n^l}\right) > 0,$$

hold. Let us fix $k > N$. If $a \neq 0$, then we define

$$g(t) = \begin{cases} f\left(q^{[k]_n}a^{n^k}\right) & \text{if } t = q^{[k+1]_n}a^{n^{k+1}} \\ 0 & \text{otherwise.} \end{cases}$$

Since $a \neq 0$, we see that

$$\int_a^b f(t)g(qt^n)\,d_{n,q}t = q^{[k]_n}a^{n^k}\left(q^{n^k}a^{n^k(n-1)} - 1\right)\left(f\left(q^{[k]_n}a^{n^k}\right)\right)^2 \neq 0,$$

which is a contradiction. If $a = 0$, then we define

$$g(t) = \begin{cases} f\left(q^{[k]_n}b^{n^k}\right) & \text{if } t = q^{[k+1]_n}b^{n^{k+1}} \\ 0 & \text{otherwise.} \end{cases}$$

Since $b \neq 0$, we obtain

$$\int_a^b f(t)g(qt^n)\,d_{n,q}t = \int_0^b f(t)g(qt^n)\,d_{n,q}t$$

$$= -q^{[k]_n b^{n^k}} \left(q^{n^k} b^{n^k(n-1)} - 1 \right) \left(f \left(q^{[k]_n} b^{n^k} \right) \right)^2 \neq 0,$$

which is a contradiction.

Let $\mathbb{E}([a, b]_{n,q}, \mathbb{R})$ be the linear space of functions $y \in \mathcal{D}([a, b]_{n,q}, \mathbb{R})$ for which the n,q-derivative is continuous and bounded on $[a, b]_{n,q}$. We equip $\mathbb{E}([a, b]_{n,q}, \mathbb{R})$ with the norm

$$\|y\|_{1,\infty} = \sup_{t \in [a,b]_{n,q}} |y(t)| + \sup_{t \in [a,b]_{n,q}} |D_{n,q}[y](t)|.$$

The following definition and lemma are similar to those of Sect. 2.1.

Definition 3.22 Let $g : [s]_{n,q} \times [-\bar{\epsilon}, \bar{\epsilon}] \to \mathbb{R}$, where

$$[s]_{n,q} := \left\{ q^{[k]_n} s^{n^k} : k \in \mathbb{N}_0 \right\}.$$

We say that $g(t, \cdot)$ is continuous in ϵ_0, uniformly in t, if and only if for every $\varepsilon > 0$ there exists $\delta > 0$ such that $|\epsilon - \epsilon_0| < \delta$ implies $|g(t, \epsilon) - g(t, \epsilon_0)| < \varepsilon$ for all $t \in [s]_{n,q}$. Furthermore, we say that $g(t, \cdot)$ is differentiable at ϵ_0, uniformly in t, if and only if for every $\varepsilon > 0$ there exists $\delta > 0$ such that $0 < |\epsilon - \epsilon_0| < \delta$ implies

$$\left| \frac{g(t, \epsilon) - g(t, \epsilon_0)}{\epsilon - \epsilon_0} - \partial_2 g(t, \epsilon_0) \right| < \varepsilon,$$

where $\partial_2 g = \frac{\partial g}{\partial \epsilon}$ for all $t \in [s]_{n,q}$.

Lemma 3.23 Let $s \in I$. Assume $g(t, \cdot)$ is differentiable at ϵ_0, uniformly in t in $[s]_{n,q}$, and that $G(\epsilon) := \int_0^s g(t, \epsilon) d_{q,\omega} t$, for ϵ near ϵ_0, and $\int_0^s \partial_2 g(t, \epsilon_0) d_{q,\omega}$ exist. Then, $G(\epsilon)$ is differentiable at ϵ_0 with $G'(\epsilon_0) = \int_0^s \partial_2 g(t, \epsilon_0) d_{n,q} t$.

Proof Without loss of generality we can assume that $s > 0$. Let $\varepsilon > 0$ be arbitrary. Since $g(t, \cdot)$ is differentiable at ϵ_0, uniformly in t, there exists $\delta > 0$ such that, for all $t \in [s]_{n,q}$ and for $0 < |\epsilon - \epsilon_0| < \delta$, the following inequality holds:

$$\left| \frac{g(t, \epsilon) - g(t, \epsilon_0)}{\epsilon - \epsilon_0} - \partial_2 g(t, \epsilon_0) \right| < \frac{\varepsilon}{s}.$$

Applying Lemmas 3.16 and 3.18, for $0 < |\epsilon - \epsilon_0| < \delta$, we have

$$\left| \frac{G(\epsilon) - G(\epsilon_0)}{\epsilon - \epsilon_0} - G'(\epsilon_0) \right|$$

$$= \left| \frac{\int_0^s g(t, \epsilon)\, d_{n,q}t - \int_0^s g(t, \epsilon_0)\, d_{n,q}t}{\epsilon - \epsilon_0} - \int_0^s \partial_2 g(t, \epsilon_0)\, d_{n,q}t \right|$$

$$= \left| \int_0^s \left[\frac{g(t, \epsilon) - g(t, \epsilon_0)}{\epsilon - \epsilon_0} - \partial_2 g(t, \epsilon_0) \right] d_{n,q}t \right|$$

$$< \int_0^s \frac{\varepsilon}{s}\, d_{n,q}t = \frac{\varepsilon}{s} \int_0^s 1\, d_{n,q}t = \varepsilon.$$

Hence, $G(\cdot)$ is differentiable at ϵ_0 and $G'(\epsilon_0) = \int_0^s \partial_2 g(t, \epsilon_0)\, d_{n,q}t$.

3.2.1 The Power Quantum Euler–Lagrange Equation

We consider the variational problem of finding minimizers (or maximizers) of a functional

$$\mathcal{L}[y] = \int_a^b f(t, y(qt^n), D_{n,q}[y](t))\, d_{n,q}t, \tag{3.11}$$

over all $y \in \mathbb{E}([a, b]_{n,q}, \mathbb{R})$ satisfying the boundary conditions

$$y(a) = \alpha, \quad y(b) = \beta, \quad \alpha, \beta \in \mathbb{R}, \tag{3.12}$$

where $f : [a, b]_{n,q} \times \mathbb{R} \times \mathbb{R} \to \mathbb{R}$ is a given function. A function $y \in \mathbb{E}([a, b]_{n,q}, \mathbb{R})$ is said to be admissible if it satisfies endpoint conditions (3.12). Let us denote by $\partial_2 f$ and $\partial_3 f$, respectively, the partial derivatives of $f(\cdot, \cdot, \cdot)$ with respect to its second and third argument. In the sequel, we assume that $(u, v) \to f(t, u, v)$ is a $C^1(\mathbb{R}^2, \mathbb{R})$ function for any $t \in [a, b]_{n,q}$ and $f(\cdot, y(\cdot), D_{n,q}y(\cdot))$, $\partial_2 f(\cdot, y(\cdot), D_{n,q}y(\cdot))$, and $\partial_3 f(\cdot, y(\cdot), D_{n,q}y(\cdot))$ are continuous and bounded on $[a, b]_{n,q}$ for all admissible functions $y(\cdot)$. We say that $p \in \mathbb{E}([a, b]_{n,q}, \mathbb{R})$ is an admissible variation for (3.11) and (3.12) if $p(a) = p(b) = 0$.

For an admissible variation p, we define function $\phi : [-\bar{\varepsilon}, \bar{\varepsilon}] \to \mathbb{R}$ by

$$\phi(\varepsilon) = \phi(\varepsilon; y, p) := \mathcal{L}[y + \varepsilon p].$$

The first variation of problem (3.11) and (3.12) is defined by

$$\delta\mathcal{L}[y, p] := \phi(0; y, p) = \phi'(0).$$

Observe that,

$$\mathcal{L}[y + \varepsilon p] = \int_a^b f(t, y(qt^n) + \varepsilon p(qt^n), D_{n,q}[y](t) + \varepsilon D_{n,q}[p](t)) \, d_{n,q}t$$

$$= \int_0^b f(t, y(qt^n) + \varepsilon p(qt^n), D_{n,q}[y](t) + \varepsilon D_{n,q}[p](t)) \, d_{n,q}t$$

$$- \int_0^a f(t, y(qt^n) + \varepsilon p(qt^n), D_{n,q}[y](t) + \varepsilon D_{n,q}[p](t)) \, d_{n,q}t.$$

Writing

$$\mathcal{L}_b[y + \varepsilon p] = \int_0^b f(t, y(qt^n) + \varepsilon p(qt^n), D_{n,q}[y](t) + \varepsilon D_{n,q}[p](t)) \, d_{n,q}t$$

and

$$\mathcal{L}_a[y + \varepsilon p] = \int_0^a f(t, y(qt^n) + \varepsilon p(qt^n), D_{n,q}[y](t) + \varepsilon D_{n,q}[p](t)) \, d_{n,q}t,$$

we have

$$\mathcal{L}[y + \varepsilon p] = \mathcal{L}_b[y + \varepsilon p] - \mathcal{L}_a[y + \varepsilon p].$$

Therefore,

$$\delta\mathcal{L}[y, p] = \delta\mathcal{L}_b[y, p] - \delta\mathcal{L}_a[y, p]. \tag{3.13}$$

Knowing (3.13), the following lemma is a direct consequence of Lemma 3.23.

Lemma 3.24 *Put* $g(t, \varepsilon) = f\left(t, y(qt^n) + \varepsilon p(qt^n), D_{n,q}[y](t) + \varepsilon D_{n,q}[p](t)\right)$ *for* $\varepsilon \in [-\bar{\varepsilon}, \bar{\varepsilon}]$. *Assume that:*

(i) $g(t, \cdot)$ *is differentiable at* 0 *uniformly in* $t \in [a, b]_{n,q}$;
(ii) $\mathcal{L}_a[y + \varepsilon p]$ *and* $\mathcal{L}_b[y + \varepsilon p]$, *for* ε *near* 0, *exist;*
(iii) $\int_0^a \partial_2 g(t, 0) \, d_{n,q}t$ *and* $\int_0^b \partial_2 g(t, 0) \, d_{n,q}t$ *exist.*

Then,

$$\delta\mathcal{L}[y, h] = \int_a^b \left[\partial_2 f(t, y(qt^n), D_{n,q}[y](t)) p(qt^n) \right.$$

$$+ \partial_3 f(t, y(qt^n), D_{n,q}[y](t)) D_{n,q}[p](t) \Big] d_{n,q} t.$$

In the sequel, we always assume, without mentioning it explicitly, that variational problems satisfy the assumptions of Lemma 3.24.

Definition 3.25 An admissible function $\tilde{y}$ is said to be a local minimizer (resp. a local maximizer) to problem (3.11) and (3.12) if there exists $\delta > 0$ such that $\mathcal{L}[\tilde{y}] \leq \mathcal{L}[y]$ (resp. $\mathcal{L}[\tilde{y}] \geq \mathcal{L}[y]$) for all admissible y with $\|y - \tilde{y}\|_{1,\infty} < \delta$.

The following result offers a necessary condition for local extremizer.

Theorem 3.26   (A necessary optimality condition for problem (3.11) **and** (3.12)**)** *Suppose that the optimal path to problem* (3.11) *and* (3.12) *exists and is given by $\tilde{y}$. Then, $\delta\mathcal{L}[\tilde{y}, p] = 0$.*

Proof   Is left to the reader. Hint: see the proof of Theorem 2.20.

From now on $\{\cdot\}$ stands for

$$\{y\}(t) := (t, y(qt^n), D_{n,q}[y](t)).$$

Theorem 3.27   (Euler–Lagrange equation for problem (3.11) **and** (3.12)**)** *Suppose that $\tilde{y}$ is an optimal path to problem* (3.11) *and* (3.12)*. Then,*

$$D_{n,q}[\partial_3 f]\{y\}(t) = \partial_2 f\{y\}(t) \tag{3.14}$$

for all $t \in [a, b]_{n,q}$.

Proof   Suppose that $\mathcal{L}$ has a local extremizer y. Consider the value of $\mathcal{L}$ at a nearby function $\tilde{y} = y + \varepsilon p$, where $\varepsilon \in \mathbb{R}$ is a small parameter, $p \in \mathbb{E}$, and $p(a) = p(b) = 0$. Let

$$\phi(\varepsilon) = \mathcal{L}[y + \varepsilon p] = \int\limits_a^b f\{y + p\}(t) \, d_{n,q} t.$$

By Theorem 3.26, a necessary condition for $\tilde{y}$ to be an extremizer is given by

$$\phi'(\varepsilon)\big|_{\varepsilon=0} = 0 \Leftrightarrow \int\limits_a^b \left[\partial_2 f\{y\}(t) p(qt^n) + \partial_3 f\{y\}(t) D_{n,q}[p](t) \right] d_{n,q} t = 0.$$

$$\tag{3.15}$$

Integration by parts (see (3.10)) gives

$$\int\limits_a^b \partial_3 f\{y\}(t) D_{n,q}[p](t) \, d_{n,q} t$$

$$= \partial_3 f\{y\}(t)p(t)|_{t=a}^{t=b} - \int_a^b D_{n,q}[\partial_3 f]\{y\}(t)p(qt^n)\, d_{n,q}t.$$

Because $p(a) = p(b) = 0$, the necessary condition (3.15) can be written as

$$0 = \int_a^b \left(\partial_2 f\{y\}(t) - D_{n,q}[\partial_3 f]\{y\}(t) \right) p(qt^n)\, d_{n,q}t$$

for all p such that $p(a) = p(b) = 0$. Thus, by Lemma 3.21, we have

$$\partial_2 f\{y\}(t) - D_{n,q}[\partial_3 f]\{y\}(t) = 0$$

for all $t \in [a, b]_{n,q}$.

Remark 3.28 Analogously to the classical calculus of variations (Weinstock 1974), to the solutions of the Euler–Lagrange equation (3.14) we call *(power quantum) extremals*.

Remark 3.29 If the function under the sign of integration f (the Lagrangian) is given by $f = f(t, y_1, \ldots, y_m, D_{n,q}[y_1], \ldots, D_{n,q}[y_m])$, then the necessary optimality condition is given by m equations similar to (3.14), one equation for each variable.

Example 3.30 Let us fix n, q, such that $1 \in I$. Consider the problem

$$\text{minimize}\quad \mathcal{L}[y] = \int_0^1 \left(y(qt^n) + \frac{1}{2}(D_{n,q}[y](t))^2 \right) d_{n,q}t \qquad (3.16)$$

subject to the boundary conditions

$$y(0) = 0, \quad y(1) = \beta, \qquad (3.17)$$

where $\beta \in \mathbb{R}$. If y is a local minimizer to problem (3.16) and (3.17), then by Theorem 3.27 it satisfies the Euler–Lagrange equation

$$D_{n,q} D_{n,q} y(t) = 1$$

for all $t \in \left\{ q^{[k]_n} : k \in \mathbb{N}_0 \right\} \cup \{0\}$. Applying Theorem 3.4 (see also Example 3.5) and Theorem 3.12, we obtain

$$y(t) = -\sum_{k=0}^{\infty} q^{[k]_n} t^{n^k} \left(q^{n^k} t^{n^k (n-1)} - 1 \right) \left(q^{[k]_n} t^{n^k} + c \right),$$

where c satisfies equation

$$\beta = -\sum_{k=0}^{\infty} \left(q^{[k+1]_n} - q^{[k]_n} \right) \left(q^{[k]_n} + c \right),$$

as the power quantum extremal to problem (3.16) and (3.17). For example, choosing
$n = 1$ and $\beta = \frac{1}{1+q}$ in (3.16) and (3.17), we get the extremal

$$y(t) = \frac{t^2}{1+q}.$$

3.2.2 Leitmann's Direct Optimization Method

Let $\bar{f} : [a, b]_{n,q} \times \mathbb{R} \times \mathbb{R} \to \mathbb{R}$. We assume that $(u, v) \to \bar{f}(t, u, v)$ is a $C^1(\mathbb{R}^2, \mathbb{R})$
function for any $t \in [a, b]_{n,q}$ and $\bar{f}(\cdot, y(\cdot), D_{n,q}y(\cdot))$, $\partial_2 \bar{f}(\cdot, y(\cdot), D_{n,q}[y](\cdot))$, and
$\partial_3 \bar{f}(\cdot, y(\cdot), D_{n,q}[y](\cdot))$ are continuous and bounded on $[a, b]_{n,q}$ for all admissible
functions $y(\cdot)$. Consider the integral

$$\bar{\mathcal{L}}[\bar{y}] = \int_a^b \bar{f}\{\bar{y}\}(t) \, d_{n,q}t.$$

Lemma 3.31 (Leitmann's power quantum fundamental lemma) *Let* $y = z(t, \bar{y})$
be a transformation having an unique inverse $\bar{y} = \bar{z}(t, y)$ *for all* $t \in [a, b]_{n,q}$ *such*
that there is a one-to-one correspondence

$$y(t) \Leftrightarrow \bar{y}(t),$$

for all functions $y \in \mathbb{E}([a, b]_{n,q}, \mathbb{R})$ *satisfying (3.12) and all functions* $\bar{y} \in$
$\mathbb{E}([a, b]_{n,q}, \mathbb{R})$ *satisfying*

$$\bar{y}(a) = \bar{z}(a, \alpha), \quad \bar{y}(b) = \bar{z}(b, \beta). \tag{3.18}$$

If the transformation $y = z(t, \bar{y})$ *is such that there exists a function* $G : [a, b]_{n,q} \times$
$\mathbb{R} \to \mathbb{R}$ *satisfying the functional identity*

$$f\{y\}(t) - \bar{f}\{\bar{y}\}(t) = D_{n,q}G(t, \bar{y}(t)), \tag{3.19}$$

then if $\bar{y}^*$ *yields the extremum of* $\bar{\mathcal{L}}$ *with* $\bar{y}^*$ *satisfying (3.18),* $y^* = z(t, \bar{y}^*)$ *yields*
the extremum of $\mathcal{L}$ *for* y^* *satisfying (3.12).*

Proof Is left to the reader. Hint: see the proof of Theorem 2.30.

Example 3.32 Let $a, b \in I$ with $a < b$, and α, β be two given reals, $\alpha \neq \beta$. We consider the following problem:

$$\text{minimize} \quad \mathcal{L}[y] = \int_a^b \left[D_{n,q}(y(t)g(t)) \right]^2 d_{n,q}t, \tag{3.20}$$

$$y(a) = \alpha, \quad y(b) = \beta,$$

where g does not vanish on the interval $[a, b]_{n,q}$. Then, the function

$$y(t) = (At + C)g^{-1}(t)$$

with $A = [\alpha g(a) - \beta g(b)] (a-b)^{-1}, C = [a\beta g(b) - b\alpha g(a)] (a-b)^{-1}$, minimizes (3.20) (Why?).

References

Abu Risha MH, Annaby MH, Ismail MEH, Mansour ZS (2007) Linear q-difference equations. Z Anal Anwend 26(4):481–494

Aldwoah KA, Malinowska AB, Torres DFM (2012) The power quantum calculus and variational problems. Dyn Contin Discrete Impuls Syst Ser B Appl Algorithms 19(1–2):93–116

Jackson FH (1908) On q-functions and a certain difference operator. Trans Roy Soc Edinburgh 46:64–72

Jackson FH (1910) On q-definite integrals. Quart J Pure and Appl Math 41:193–203

Kac V, Cheung P (2002) Quantum calculus. Springer, New York

Weinstock R (1974) Calculus of variations. With applications to physics and engineering. Dover, New York (Reprint of the 1952 edition)

Chapter 4
Conclusion

In this small book we consider variational problems in the context of the Hahn quantum calculus. Such variational problems are defined through the Hahn quantum difference operator and the

Jackson–Nörlund integral (Hahn 1949; Jackson 1910, 1951; Nörlund 1924). The origin of the Hahn quantum difference operator dates back to a 1949 paper of Hahn (1949), where it was introduced to unify, in a limiting sense, the Jackson q-difference derivative and the forward difference. For both of these latter two quantum difference operators, variational problems have been studied previously. The forward difference problems were studied at least as early as 1937 by Fort (1937) and for the q-difference by Bangerezako (2004). In both of these works the authors discuss necessary conditions for optimality and obtain the analogue of the classical Euler–Lagrange equation, as well as other classical results. Our main goal was to present, in a unified way, extensions of these results to the Hahn quantum difference operator.

Another related and interesting course of study is that of the notion of time scale. The origins of this idea dates back to the late 1980s when S. Hilger introduced the concept in his Ph.D. thesis (directed by B. Aulbach) and showed how to unify continuous time and discrete time dynamical systems (Aulbach and Hilger 1990; Hilger 1990). Since Hilger's seminal work, the literature has exploded with papers and books on time scales in which many known results for ordinary differential equations and difference equations have been combined and extended (Bohner and Peterson 2001; Malinowska and Torres 2009, 2010; Martins and Torres 2011a; Mozyrska et al. 2010). The classical results of the calculus of variations were first extended to times scales by Bohner (2004) and then developed by several different authors: see Atici and McMahan (2009); Cresson et al. (2012); Malinowska and Torres (2011), Martins and Torres (2011b, 2012) and references therein. However, the Euler–Lagrange equations here obtained are not comparable with those of time scales. Indeed, the Hahn quantum calculus is not covered by the Hilger time scale theory. This is well explained, for example, in the 2009 Ph.D. thesis of Aldwoah (Aldwoah

2009); see also (Aldwoah and Hamza 2011). Here we just note the following: if in the time-scale calculus one chooses the time scale to be the q-scale $\mathbb{T} := \{q^n : n \in \mathbb{Z}\}$, then the expression of the delta-derivative coincides with the expression of the Jackson q-difference derivative. However, they are not the same. There is an important distinction: the Jackson q-difference derivative is defined in the set of real numbers (Kac and Cheung 2002), while the time-scale derivative is only defined in the subset $\mathbb{T}$ of the real numbers (Bohner and Peterson 2001). One more difference, between the Hahn calculus we use here and the time scale theory, is the following: the delta integral satisfies all the usual properties of the Riemann integral while this is not true for the Jackson–Nörlund integral. Indeed, the inequality of Remark 2.13 does not always hold for the Jackson–Nörlund integral.

The main advantage of our results is that they are able to deal with nondifferentiable functions, even discontinuous functions, that are important in physical systems. Quantum derivatives and integrals play a leading role in the understanding of complex physical systems. For example, in 1992 Nottale introduced the theory of scale-relativity without the hypothesis of space-time differentiability (Nottale 1992). A rigorous mathematical foundation to Nottale's scale-relativity theory is nowadays given by means of a quantum calculus (Almeida and Torres 2009; Cresson et al. 2009). We remark that results in time scales are not able to deal with such nondifferentiable functions. Variational problems on time scales are formulated for functions that are delta (or nabla) differentiable and, as it is well known, time-scale differentiability implies continuity. This is not the case in our context: see Example 2.2, where a discontinuous function is q, ω-differentiable in all the real interval $[-1, 1]$.

We believe that the obtained results are also of interest in Economics. Economists model time as continuous or discrete. The kind of "time" (continuous or discrete) to be used in the construction of dynamic models is a moot question. Although individual economic decisions are generally made at discrete time intervals, it is difficult to believe that they are perfectly synchronized as postulated by discrete models. The usual assumption that the economic activity takes place continuously, is a convenient abstraction in many applications. In others, such as the ones studied in financial-market equilibrium, the assumption of continuous trading corresponds closely to reality. We believe that our Hahn's approach helps to bridge the gap between two families of models: continuous and discrete. We trust that the field here initiated will prove fruitful for further research.

References

Aldwoah KA (2009) Generalized time scales and associated difference equations. Ph.D. Thesis, University of Cairo, Cairo

Aldwoah KA, Hamza AE (2011) Difference time scales. Int J Math Stat 9(A11):106–125

Almeida R, Torres DFM (2009) Hölderian variational problems subject to integral constraints. J Math Anal Appl 359(2):674–681

Atici FM, McMahan CS (2009) A comparison in the theory of calculus of variations on time scales with an application to the Ramsey model. Nonlinear Dyn Syst Theory 9(1):1–10

Aulbach B, Hilger S (1990) A unified approach to continuous and discrete dynamics. In: Qualitative theory of differential equations (Szeged, 1988), Colloq Math Soc János Bolyai, 53, North-Holland, Amsterdam, pp 37–56

Bangerezako G (2004) Variational q-calculus. J Math Anal Appl 289(2):650–665

Bohner M (2004) Calculus of variations on time scales. Dyn Syst Appl 13(3–4):339–349

Bohner M, Peterson A (2001) Dynamic equations on time scales. An introduction with applications, Birkhäuser, Boston

Cresson J, Frederico GSF, Torres DFM (2009) Constants of motion for non-differentiable quantum variational problems. Topol Methods Nonlinear Anal 33(2):217–231

Cresson J, Malinowska AB, Torres DFM (2012) Time scale differential, integral, and variational embeddings of Lagrangian systems. Comput Math Appl 64(7):2294–2301

Fort T (1937) The calculus of variations applied to Nörlund's sum. Bull Am Math Soc 43(12):885–887

Hahn W (1949) Über orthogonalpolynome, die q-differenzenlgleichungen genügen. Math Nachr 2:4–34

Hilger S (1990) Analysis on measure chains–a unified approach to continuous and discrete calculus. Results Math 18(1–2):18–56

Jackson FH (1910) On q-definite integrals. Quart J Pure Appl Math 41:193–203

Jackson FH (1951) Basic integration. Quart J Math Oxford Ser 2(2):1–16

Kac V, Cheung P (2002) Quantum calculus. Springer, New York

Malinowska AB, Torres DFM (2009) On the diamond-alpha Riemann integral and mean value theorems on time scales. Dyn Syst Appl 18(3–4):469–481

Malinowska AB, Torres DFM (2010) Leitmann's direct method of optimization for absolute extrema of certain problems of the calculus of variations on time scales. Appl Math Comput 217(3):1158–1162

Malinowska AB, Torres DFM (2011) Euler-Lagrange equations for composition functionals in calculus of variations on time scales. Discrete Cont Dyn Syst 29(2):577–593

Martins N, Torres DFM (2011a) L'Hôpital-type rules for monotonicity with application to quantum calculus. Int J Math Comput 10(M11):99–106.

Martins N, Torres DFM (2011b) Generalizing the variational theory on time scales to include the delta indefinite integral. Comput Math Appl 61(9):2424–2435

Martins N, Torres DFM (2012) Necessary optimality conditions for higher-order infinite horizon variational problems on time scales. J Optim Theory Appl 155(2):453–476

Mozyrska, D, Pawłuszewicz, E, Torres, DFM (2010) The Riemann-Stieltjes integral on time scales. Aust J Math Anal Appl 7(1): 14

Nörlund N (1924) Vorlesungen über Differencenrechnung. Springer Verlag, Berlin

Nottale L (1992) The theory of scale relativity. Internat J Mod Phys A 7(20):4899–4936

Index

A. B. Malinowska and D. F. M. Torres, *Quantum Variational Calculus*,
SpringerBriefs in Control, Automation and Robotics,
DOI: 10.1007/978-3-319-02747-0, © The Author(s) 2014